THE IMPULSE OF RENEWAL FOR CULTURE AND SCIENCE

A lecture series by
Rudolf Steiner

7 lectures given in Berlin
between 6-11 March 1922 to the High School.

Translated by Hanna von Maltitz

Cover artwork by Hanna von Maltitz
Her painting is entitled,
"Lost Souls"

Visit Hanna's works on-line at Fine Art Presentations:
fineart.elib.com/fineart.php?dir=Site_index/Anthroposophical/Maltitz_Hanna_von

Cover designed by James D. Stewart

Rudolf Steiner
Visit the website at https://www.rsarchive.org/

Printed in the United States of America

First Printing: September 2018
Rudolf Steiner Archive & e.Lib

ISBN-13: 978-1-948302-04-3
ISBN-10: 1-948302-04-7

CONTENTS

"ANTHROPOSOPHY AND NATURAL SCIENCE."

Lecture One by Rudolf Steiner given in Berlin, 6 March 1922

Welcome, all who are present here! It was the wish of the committees of this High School week that I give an introduction each day regarding the course which will take place in a scientifically orientated process. It will be conducted with the aim of Anthroposophy fructifying the individual branches of science and of life, and with these introductory words I ask you to take up this first lecture.

What has surprised me the most at the reception of the anthroposophical research method is the opposition, particularly from the philosophic-natural scientific side — I'm not only saying the natural scientific side — brought against Anthroposophy, and it stems from a basic belief that Anthroposophy's methods stand in an unauthorised, opposing position to those of natural science which has developed exponentially in the last century, particularly the 19th Century. It seems to me that among all the various things related to Anthroposophy which our contemporaries find the most difficult to understand, is this, that Anthroposophy in relation to natural science doesn't want anything other than that the methods used by natural science which have proved so fruitful, be developed further in a corresponding manner. In any case, with the idea of further development something else needs understanding, if one wants to arrive at an anthroposophic understanding, than that which one usually calls further evolution from a theoretical point of view.

Further development from a theoretical point of view for most people means that the particular way thoughts are linked — particularly if I may express it as the field of thought — remains constant, also when relevant thought systems expand to other areas of the world's phenomena. So for instance when you engage scientifically with lifeless, inorganic nature you necessarily come to linking thoughts, to a certain field of thought, which means the sum of linking thoughts is a foundation, in order to gradually arrive at a theory about lifeless, inorganic nature. This system of thoughts, as it stands, you now want to extend when you enter another sphere of the world, for example the sphere of organic phenomena in nature, in order to understand it. You would want with this causal orientation, which has proved itself so fruitful in the inorganic area, to simply apply it to living beings and in the same terms, drenching and explaining it, thus gradually conceptualising the sphere of the living beings similarly into an effective system derived from inorganic causalities which you would be doing with regard to lifeless, inorganic nature. What you have appropriated as a system of thought derived from lifeless nature, you simply apply to organic nature. This is what is usually understood today, as the 'expansion' of thoughts and theories.

This is of course quite the opposite of what Anthroposophy regards under such an idea as the expansion of thinking. A fully rounded concept of an independently developed, self-metamorphosed idea need to be contained, so that if you want to go from one sphere of world phenomena into another, that you don't merely apply what you have learnt from lifeless natural phenomena, and — I could call it "logically apply" — it on to life-filled phenomena in nature. By comparison, just as things change in the living world, growing, going through metamorphosis, and how they often become unrecognisable from one form into another, so thoughts should also take on other forms when they enter into other spheres. One thing remains the same in all spheres which is what gives the scientific point of view its monistic and methodical character, it's the manner and way in which you can position yourself internally to what can be called "scientific certainty" which forms the basis of scientific convictions.

Whoever wants to find proof why one can't use concepts gleaned from lifeless nature, concepts which are applied through habits, in which to verify human causalities — if I may use _Du Bois-Reymond_'s expression — whoever gets to know this intimately, can then shift over to quite different concepts, concepts which are metamorphosed from earlier concepts, and sound convincing in the world of the living. The way in which the human being is positioned within the scientific movement is completely monistic right though the entire scientific world view. This is usually misunderstood and results in the anthroposophic-scientific viewpoint not having a monistic but a dualistic character.

The second item which commonly leads to misunderstandings is phenomenology, to which Anthroposophy with regard to natural science must submit. We are experiencing a fruitful time of scientific developments, a time in which the important scientific researcher _Virchow_ gave his lecture regarding the separation of the philosophic world view from that of the natural scientific view, how everything had been conquered which at that time had a certain historic rating of fruitful concepts regarding the inorganic, resulting in a certain rationalism being established in science. This period which worked on the one side earnestly from empiricism against the outer world of facts, this still went over to a far-reaching rationalism when it came down to it to elucidate the empirically explored facts of nature.

By contrast we now have the standpoint of Anthroposophy which comes from — at least for me it comes from this, if I might make a personal remark — from the Goethean conception of nature. Anthroposophy stands on the basis of a phenomenological concept of nature. In a certain way this phenomenology of recent times was established again by <u>Ernst Mach</u>, and as he established it, he again appeared to reveal fertile points of view, if one complies with his boundaries. For <u>Goethe</u> it simply lies in his words: The world of appearances is theory enough, one doesn't even need to subscribe to artificial theories. The blue of the sky is a phenomenon which stops there, and one can't condescend in a rational way through mere thoughts behind the appearance by looking for hypothetical, assumed reasons for explanations. Goethe arrived at this point by establishing what he called the 'Ur-phenomenon' ("Urphänomen" or 'Original Phenomenon'). It is self-explanatory that in the course of the fertile time for science in the 19th Century, much of which has become obsolete, what Goethe envisaged for natural science was the following: The methodical, the way of thinking itself which Goethe introduced into natural science is not only overhauled today but it appears to me to be not sufficiently understood.

I know very well how in the 19th Century several — one could say nearly all — of the details of Goethe's interpretations regarding natural scientific things have been overhauled. Despite that, I would like to sustain a sentence today which I made in the eighties of the previous (19th) Century in relation to Goethe's concept of nature: 'Goethe is the Copernicus and Kepler of organic natural science.' I want to still support this sentence today because I believe the following is justified by it.

What is it that lets us finally arrive at a true perception of nature on which so much of the 19th Century had been achieved? What I'm referring to can't but be set within the boundary of a historic category. What has been achieved through science during the 19th century nearly always refers back to the application of mathematical methods because even where pure mathematics aren't applied, but thoughts steered according to other principles of causality, where theories are developed, here the mathematical way of thinking forms a basis.

It is significant in what happened: we have seen in the course of the 19th Century how certain parties of science in a certain rationalistic way had to form a foundation by the introduction of mathematics. The Kantian saying claims that there is only as much certainty in a science as there is mathematics contained within it. Now obviously mathematics can be introduced into everything. Claims of causality go further than possible mathematical developments of concepts. However, what has been done in terms of explanations of causality was done extensively according to the pattern of mathematical conceptions. When Ernst Mach became involved, considering it with his more phenomenological viewpoint of these concepts of causality, as it had developed in the course of the 19th Century, he wanted to arrive at a certain causal understanding of the contents. Finally he declared: 'When I consider a process and its cause, there is actually nothing different between it and a mathematical function. For instance, if I say: X equals Y, while X is the cause under the influence of the working called Y, then I have taken the entire thing back to the concept which I have in mathematics, when I created a concept of function. It can also be seen in the history of science, how the concept of mathematics has been brought into the sphere of science.'

Now Goethe is usually regarded — with a certain right — as a non-mathematician; he even called himself as such. However, if one places Goethe there as a non-mathematician, then misunderstandings arise — somewhat in the sense that Goethe couldn't achieve much with mathematical details, that he was not particularly talented in his time to solve mathematical problems. That may of course be admitted. I also don't believe Goethe in his total being had particular patience to solve detailed mathematical examples, if it was more algebraic. That has to be admitted. However, Goethe had in a certain sense, as paradoxical as it might sound, more of a mathematician's brain than some mathematicians; because he had fine insight into mathematized nature, in the nature of building mathematical concepts, and he prized this way of thinking, which lives within the soul process also with the content of imagination when concepts are created.

The mathematician, when building concepts, scrutinizes everything internally. Take for instance a simple example of Euclidean geometry which proves that the three angles of a triangle amounts to 180 degrees, where, by drawing a line parallel to the base line, through the tip of the triangle, two angles are created, which are equal to the other two angles in the triangle — the angle in the tip remains the same of course — and how one can see that these three corners at the top add to 180 degrees, being the total of all the corners of the triangle. When you consider this, you can see that with a mathematical proof you have simultaneously something which is not dependent on outer perception, but it is completely observed as an inner creation. If you then have an outer triangle you will find that the outer facts can be verified with one's previous inner scrutiny. That is so with all mathematics. Everything remains the same, no perception of the senses need to be added to it in order to arrive at what is called a "proof", that everything which has been discovered internally can be verified, piece-by-piece.

It is this particular kind of mathematics which Goethe regarded as eminently scientific and insofar he actually had a good mathematical brain. This for example also leads to the basis of the famous lectures which Goethe and Schiller, during the time of their blossoming friendship, had led regarding the method of scientific consideration. They had both attended a lecture held by the researcher *Batsch* in the Jena based Naturforschenden Gesellschaft (Nature's

Investigator's Club — Wikipedia: August J G Batsch). As they departed from the lecture, Schiller said to Goethe that the content of what they had heard was a very fragmentary method of observing nature, it didn't bring one to a whole. — One can imagine that Batsch simply took single natural objects and ordered them one below another and refrained, as befitted most researchers at the time, from ordering them somehow which could lead to an overall view of nature. Schiller found this unsatisfactory and told Goethe. Goethe said he understood how a certain unification, a certain wholeness had to be brought into observations of nature. Thus, he began with a few lines — he narrates this himself — to draw the "Urpflanze" (Original Plant), how it can be thought about, looked at inwardly — not like some or other plant encountered in the day, but how it could be regarded inwardly through the root, stem, leaves, flowers and fruit.

In my introduction to Goethe's "Naturwissenschaftlichen Schriften" (Natural Scientific Notes) of the 80's of the previous (19th) century, I tried to copy the image which Goethe presented to Schiller on paper. — Schiller looked at it and said, as was his way of expressing himself: 'This is no experience, this is an idea.' Schiller actually meant that if one made a drawing of something like that, it had been spun out of oneself, it is good as an idea and as a thought but in reality, it has no source. Goethe couldn't understand this way of thinking at all, and finally the conversation was concluded by his reply: 'If that is the case then I can see my ideas with my eyes.'

What did Goethe mean by this? He meant — but hadn't expressed it like this, he meant: 'When I draw a triangle its corners add up to 180 degrees by themselves; when I have seen as many triangles and constructed them within me, the sum of all triangles fit on to this triangle, I have in this way gained something from within which fits the totality of my experiences.' In this way Goethe wanted to draw his "Urpflanze" according to the "Ur-triangle" and this Ur-plant would have such characteristics that one could find it in all individual plants. Just as the sum of the triangle's corners, when you draw the Ur-triangle, amounts to 180 degrees, so also this ideal image of the Ur-plant would be rediscovered in each plant if you go through an entire row of plants.

In this manner Goethe wanted all of science to take shape. Essentially he wanted — but he couldn't continue — to let organic science be developed and introduce such methods of thinking as had been proven so fruitful for inorganic science. One can very clearly see, when Goethe writes about Italy, how he developed the idea of the Ur-plant ever further. He more or less said: 'Here among the plants in South Italy and Sicily in the multiplicity of the plant world the Ur-plant rose up for me specially, and it must surely find an image which all actual plants possibly have within them, an image in which many different sides may appear taking on this or that, adapting elongated or other plant forms, soon forming the flower, soon the fruit and more, and so on — just like a triangle can have sharp or blunt corners.' Goethe searched for an image according to which all plants could be formed. It is quite incorrect when later, *Schleiden* [Matthias Jakob Schleiden (1804-1881), botanist, Physician and lawyer. — "The plant and Its Life", 6th edition of Leipzig 1864, Lecture 4: "The Morphology of Plants", p. 86: "The idea of such laws for the design of the plant was first developed by Goethe in his idea of 'Urpflanze', what he put forth as the primal, or ideal plant. That realization was, as it were, the task of nature, and which she more or less has completely achieved completely."] indicated that Goethe was looking for an actual plant to fit his Ur-plant. This is not correct — just as when a mathematician, when speaking about a triangle, doesn't have a particular triangle in mind — so Goethe was referring to an image, which, proven inwardly, could actually be verified everywhere in the outer world.

Goethe basically had a mathematical brain, much more mathematical than those who develop Astronomy. That's the essence. This led to Goethe, in his conversation with Schiller, to say: 'Then I see my ideas with my eyes.' He saw them with his eyes because he could pursue them everywhere in the phenomena. He didn't go along with anything only being an "idea", because he found complete resonance in the experience of building an idea; just like a mathematician senses harmony within the experience of creating mathematical ideas. This led Goethe, if I might say so, through an inner consequence to arrive at mere phenomenology, that means not trying to find anything behind appearances as such, above all not to create a rationalistic world of atoms.

Here one enters into the area where many — I can but say it — misunderstandings developed relating to battles against some scientific philosophic points of view. It simply meant that what the outer world offered the senses were seen as phenomena. Goethe and with him the entire scientific phenomenology was narrowed down to not going directly from some sense perceptive phenomenon into the atomic content behind it, but by focussing *purely* on the perceived phenomenon and the single element of the perceived facts, and then to search not for what lies behind it, but for its correlation to other elements of sense-perceptible appearances relating to it.

It is very easy — I understand totally where misunderstandings come from — to find such phenomenology as hopeless. One can say for instance: When one wants to merely narrow down descriptions of mutual relationships in sensory phenomena and search for those phenomena which are the simplest, which possibly have the most manageable facts — which Goethe calls "Ur-phenomena" — then one doesn't come to an observation about endless fruitful things as modern Chemistry has delivered for example. How — one could ask — can one actually arrive at atomic weight ratios without observing the atomic world? Now, in this case one can counter this with the question: When one really reflects what is present there, does it involve a need to start with the phenomenon? One has no involvement with it. With atomic weight ratios one is involved with phenomena, namely weight ratios. Still, one could ask: To go further, could one express the atomic weight ratios numerically in order to clarify how specific molecular structures are built out of pure thought, rationalistically? One could pose this question as well. Briefly, what is not involved when Goethean observation is used, is this: remaining stuck in the phenomena themselves. I would like to compare it with a trivial comparison.

Let's imagine someone is confronted with a written word. What will he do? If he hasn't learnt to read, he would meet it as something inexplicable. If he was literate, he would unconsciously join the single forms together and encounter its meaning within his soul. He certainly wouldn't start with each symbol, for instance by taking the W and search for its meaning, by approaching the upward stroke, followed by the descending stroke, in order to discover the foundation of the letters. No, he would read — and not search for the underlying to obtain clarity. In this way phenomenology wants to "read" as well. You may remain within the connections of phenomenology and learn to read them, and not, when I offer a complexity of phenomenon, turn back to atomic structures.

It comes down to entering into the field of phenomena and learning to read within their inner meaning. This would lead to a science which has nothing rationalistically construed within it behind the phenomena, but which, simply through the way the phenomena are regarded, lead to certain legitimate structures. In every case this science would be a member of the totality of the phenomena. One would speak in a specific way about nature. With this approach the laws of nature would be contained, but in every instance the phenomena themselves would be contained in the forms of expression. One would achieve what I would like to call a natural science inherent in the phenomena. Along the lines of such a science was Goethe's striving. The way and manner of his approach had to be changed according to the progress of modern times, but it still is possible for the fundamental principles to be maintained. When these fundamental principles are adhered to, nature itself presents something towards human conceptualising, which I would like to characterise in the following way.

It is quite obvious that we as modern humanity have developed our scientific concepts according to inorganic nature. This is the result of inorganic natural phenomena being relatively simple; it was also the result of, or course, when one enters the organic realm, the agents of the lifeless processes still persist. When one moves from the mineral to the plant kingdom then it does not happen that the lifeless activities stops in the plant; they only become absorbed into a higher principle, but it continues in the plant. We do the right thing when we follow the physical and chemical processes further in the plant organism according to the same point of view which we are used to following in inorganic nature. We also need to have the ability to shift our belief system towards change, to metamorphosed concepts. We need to research how the inorganic also applies to the plants and how the same processes which are found in lifeless nature, also penetrate the plants. However, this could result in the temptation to only research what lies in the mineral world within the plant and animal and as a result overlook what appears in higher spheres. Due to special circumstances this temptation increased much more in the course of the 19th Century. This happened in the following way.

When one looks at lifeless nature one feels to some extent satisfied because research of the phenomena can be done with mathematical thinking. It is quite understandable that Du Bois-Reymond in his wordy and brilliant manner gave his lecture "Regarding the boundaries of Nature's understanding" in which he, I could call it, celebrated the Laplace world view and called it the "astronomical conception" of the entire natural world existence. According to this astronomical conception not only were the starry heavens to be regarded this way, through mathematical thinking constructing single phenomena into a whole, as far as possible, but that one should try and penetrate with this into the constitution of matter. One molecule was to construct a small world system where the atoms would move in relation to one another like the stars in the world's structure. Man constructed himself in the smallest of the small world system and was satisfied that he would find the same laws in the small as in the big. So one had in the single atoms and molecules a system of moving bodies like one has outside in the world structure's system of fixed stars and planets. This is characteristic of the direction in which mankind was striving particularly in the 19th Century and how people were satisfied, as Du Bois-Reymond said, as a result of the need for causality. It simply developed out of the urge to apply mathematics fruitfully to all-natural phenomena. This resulted in the temptation for these mathematicians to remain stuck in their observation of natural phenomena.

It won't occur to anyone, also not an Anthroposophist, if he doesn't want to express himself inexpertly, to deny that this is justified, for instance when someone remains within the phenomena and concerns himself with details, for example in Astronomy, and conceive it in this way. It won't occur to anyone to start a fight against this. However, in the course of the 19th Century it occurred that everything the world offered was overlooked which had a *qualitative* aspect and only regarded the qualitative aspect by applying mathematical understanding to it. Here one must differentiate: One can admit that this mechanical explanation of the world is valid, nothing can be brought against it. One needs to differentiate between whether it can be applied justifiably to certain areas only, or whether it should be applied as the one and only possible system of understanding everything in the world.

Here lies the point of difference. The Anthroposophist will not argue in the least against something which is justified. Anthroposophy namely won't oppose the other and it is interesting to follow arguments how Anthroposophy actually admits to all which is within justifiable boundaries. It doesn't occur to the Anthroposophist to argue against what natural science has validated. However, it comes down to whether it is justified to include the entire sphere of phenomena with the mathematical-causal way of thinking, or whether it is justifiable, out of the totality of phenomena, to place those of a purely mathematical-causal abstraction as a "conceived" content, as it had been done in earlier atomic theory.

Today atomic theory has to a certain extent become phenomenological, and to this extent Anthroposophy concurs. However even today it comes down to some spooks of the 19th Century appearing in this un-Goethean atomic theory, which doesn't limit itself to phenomena but constructs a purely conceptual framework behind the phenomena. When one isn't clear about it being a purely conceptual framework, that the world searches behind phenomena, but that the appearance claims this conceptual framework is reality, then one becomes nailed down by it. It is extraordinary how such conceptual frameworks nail people down. Through them they become more dogmatic and say:

'There are people who want to explain the organic through quite different concepts which they find from quite somewhere else, but this doesn't exist; we have developed such conceptual structures which encompass the world behind the phenomena; this is the only world and this must also be the only workable way with regard the organic sphere.' — In this way the observation of the organic sphere is imported into the phenomena observed in inorganic nature; the organic is seen as having been created in the same way as inorganic nature.

Here clarity needs to be established. Without clarity no real foundation for a discussion can be created. Anthroposophy never intends sinning against legitimate methods in a dilettante manner, it will not sin against justified atomic theory; it wants to keep the route free from the creation of thought structures which had been developed earlier for the inorganic sphere and now needs to be created for other areas of nature. This will happen if one says to oneself: 'In the phenomena I only want to "read", that means, what I finally get out of the content of natural laws, dwell within the phenomena themselves — just as by reading a word, the meaning is revealed from within the letters. If I lovingly remain standing within the phenomena and am not intent on applying some hypothetical thought structure to it, then I would remain free in my scientific sense for the further development of the concept.' This ability to remain free is what we need to develop.

We may not take a system of beliefs which have been fully developed and nailed down for a specific area of nature and apply it to other areas. If we develop mere phenomenology which can obviously only happen if one takes the observed, or through an experiment of chosen phenomena which have been penetrated with thought and is thus linked to natural laws, one remains stuck within the phenomena, but now one arrives at another kind of relationship to thoughts themselves; one comes to the experience of how phenomena already exist within the laws of nature and how they now appear in our thoughts. If we allow ourselves to enter into these thoughts we no longer have the justification in as far as we are remaining within the phenomena, to speak of subjective thoughts or objective laws of nature. We simply dive into the phenomena and then give thought content to the content of the natural laws, which presents us with the things themselves. This is how Goethe could say naively: 'Then I see my ideas in Nature' — which were actually laws of nature — 'with my own eyes.'

When you position yourself in *this* particular way in the phenomena of inorganic nature, then it is possible to go over into the organic, also within scientific terms. When a person sees that his horse is brown or a gray (Schimmel) horse is white, he won't refer back to the inorganic colour but refer to what is living in a soul-spiritual way in the organism itself. People will learn to understand how the empowered inner organisation of the animal or plant produces the colour out of themselves.

In addition, it is obvious that all the minutiae, for example the functioning of metabolism, need to be examined from within. However, then one doesn't apply the organic to what one has found in the inorganic. One doesn't nail oneself firmly on to a specific system of thought, and one doesn't apply the same basic conviction you had in one area on to other areas. One remains more of a "mathematical mind" than those who refuse to allow concepts to metamorphose into the qualitative. Then one is able to reach higher areas of nature's existence through inner examination just as one is able to validate through inner examination, the lifeless mathematical structures. This is what I briefly wanted to sketch for you, and if it is expanded further, will show that the scientific side of Anthroposophy is always able, what Goethe calls being accountable, to all, even the most diligent mathematician. This was Goethe's goal with the development of his idea of the Ur-plant, which he came to, and the idea of the Ur-animal, at which he didn't arrive. Anthroposophy strives to allow the origins of Goethe's world view to emerge with regards to nature's phenomena and from the grasping of the vital element in imagination to let it rise to the form of the plant and to the form of the animal. Already during the eighties (1880's), I indicated that we need to metamorphose concepts taken from inorganic for organic nature. I'll speak more about this during the coming days.

As a result of this one comes to perceive within the organic what the actual principle of the process, the formative principle, is. Now, in conclusion of this reflection I would like to introduce something which will lead to further observations in the coming days; something which will show how this materialistic phase of scientific development is not be undervalued by Anthroposophy.

Anthroposophy must see an important evolutionary principle in this materialistic phase of natural science, an educational method through which one has once learnt to submit oneself to the empiricism of the outer senses. This was extraordinarily educational for the development of mankind, and now when this education has been enjoyed, one can look at certain things with great clarity. Whoever now, equipped with such a scientific sense for observing the outer material world, will make the observation that this material world is 'mirrored' in people, if I might use this expression.

The world we experience within is more or less an abstraction of an inner image permeated by experiences and will impulses of the outer material world so that when we move from the material outer world to the soul-spiritual, we come to nothing but imagery. Let's hold on to this firmly: outwardly there's the totality of material phenomena, which we are looking at in a phenomenological sense — and within, the soul-spiritual which has a particular abstract character, a pictorial character. If one approaches the observation with an anthroposophical view that the spiritual lies at the basis of the outer material world, the spirit which works in the movement of the stars, in the creation of minerals, plants and animals, then one enters in the spiritual creation of the outer world; one gets to know this through imagination, inspiration and intuition, then this is also an inner mirror image of the human being. But what is this inner mirror image of the human being? It is our physical organs. They respond to me in what I've learnt to know as the nature of the sun, the nature of the moon, minerals, plants, animals and so on; this is how the inner organs answer me. I only get to know my inner human organism when I get to know the outer things of the world. The material world outside mirrors in my soul-spiritual; the soul spiritual world outside reflects itself in the form of lungs, liver, heart and so on. The inner organs are, when you look at them, in the same relationship with the spiritual outer world as the relationship of our thoughts and experiences are to the material outer world.

This shows us how Anthroposophy consistently does not want to reject materialism in an enthusiastic sense. Look at the entire scope of natural science: thousands will be dissatisfied with results obtained through the usual methods of natural science. Anthroposophy and its methods will gradually gain an opinion regarding the material world which does not result in dissatisfaction. It acknowledges matter in its own organisation and in the phenomenology of the environment, but it has to acknowledge at the same time that the inner organisation is the result, the consequences of the cosmic soul-spiritual. Through this it wants to supplement what has only mathematically been accomplished in astronomy, astrophysics, physics or chemistry. This it wants to explore further in an organic cosmology and so on, and as a result bring about an understanding with materialistic people. In this lies the foundation of what Anthroposophy wants to offer to medicine, biology and so on.

So I believe that through these indications which I've only been able to give as a sketch, it will point out how Anthroposophy, when it is correctly understood, can't be seen as wanting to initiate a war against today's science but on the contrary, that the present day representatives of science haven't crossed the bridge to Anthroposophy to see how it also wants to be strictly scientific with regards to natural phenomena.

"THE HUMAN AND THE ANIMAL ORGANISATION."

Lecture Two by Rudolf Steiner given in Berlin, 6 March 1922

Welcome, all who are present here! In this second lecture I would like you to consider that I had assumed last night, to hearing Dr Kolisko's lecture today, and not present it myself. Due to this it wasn't possible in this short time to quite sort out what I would say, and I can only hope, as a whole, to roughly cover the details which Dr Kolisko wanted to convey to you.

When from the anthroposophical viewpoint the relationship of the animal world to the human world is spoken about, then it must be pointed out in particular how the present anthroposophical ideas relate historically to the Goethean world view — I have mentioned this twice here at least. The theme in question today, specifically the very first of Goethe's accomplishments in the natural scientific area, will come under scrutiny, namely in his treatise entitled: "Human beings, like the animals, are attributed with an intermaxillary jawbone in the upper jaw." One needs to keep all the relationships in view when considering how Goethe came to this treatise on the basis of some anatomical and physiological studies and on which grounds of his approach to embryological studies he attributed this to.

During the time when Goethe, already as a young student and later as a friend who to a certain extent had made the Jena University Institute dependent on him, lived with these problems to which he was exposed, namely the problem of what the actual difference could be between human beings and animals. He noticed how people all around him were focused on discovering the difference within the form, within the human and animal morphology, of the differentiation between people, who should be, to some degree, the crown of creation, and the animal world. Also, regarding the circumstances where the intermaxillary, which is clearly detached in all animals from any other jaw bone but which is not found in the human being as a separated bone, made people believe that this part of the head's development gave the decisive difference between humans and animals. Goethe didn't agree. He was of the opinion that man and animal were created according to their entire organisation in the same way, therefore such a detail could not indicate a differentiation. In addition, the intermaxillary bone in mature people grow together, so Goethe tried to show how this phenomenon relied only on later development because in the embryonic stage, human beings displayed the same relationship in their upper intermaxillary bone as in animals. You only have to follow the enthusiasm with which Goethe pointed out to how lucky he was, that the human being actually has the intermaxillary jaw bone in common with the animals, and how out of the whole big picture of the morphology, no decisive difference between the human being and the animal could be found in a single detail. From the kind of limitation of man and animal as you find everywhere in the 18th Century, it could not be stated in this way — also for Anthroposophy it could not be stated in this way. What Goethe accepted was this: By the animal organisation developing up into the human organisation, details already in the animal organ formation were transformed and then gradually through its evolution created the possibility to make space for what is within man, and in such a way reveal the transformed animalistic organisation in the totality of man. Goethe thought only about the metamorphosed animal organisation within the human and not of an autonomously separated human morphology.

This, I might add, needs to be established as the foundation in the search for the differentiation, in the anthroposophical sense, between animal and human organisms. If an organisation itself, in its forms, only depends on the animal and the human metamorphosis, then one has to, if you are looking for differentiation, primarily watch the course of life of a person and of an animal, and gradually observe how the human being unfolds out of the functioning of his organs, and how an animal takes on form through his organs. In brief, one needs to search more from biological than from morphological regions.

Now, one can prepare, in a specific way, how to discover the understanding through biological differences, by finding a foundation in which the animal functions originate and which appears in both man and animal, and this relates to the sense organs. The sense organs, or better said, the functions of the sense organs are more or less vital in everything which takes place in the animal and human organisms. We may assume that in the simple nutritional processes in the lower animals, in the purely digestive processes, a function of a primitive sense happens, which, we say, is where taste experiences for instance happen more or less as a purely chemical function of metabolism. These things become increasingly differentiated the further one ascends the animal row, right up to the human being. We won't get anywhere if we go straight to the animal organisation to find something which does not have a sensory life. Certainly one could say: what for instance do the senses have to do with the development of the lymph and blood?

Today already there is talk from the non-anthroposophical scientists about subconscious processes in the human psyche, and as a result we will for the sake of brevity only hint at it, and not let it appear as something completely unauthorised when I say: What takes place in the mouth and palate as a taste experience, what takes place in the process and function with for instance the ptyalin, pepsin and so on, how can it not also take part in the subconscious? Why should — I say this as a kind of postulation — the experience of taste not continue through the entire organism and why should the subconscious experience of taste not happen parallel to the lymph and blood development in all organ processes? Through this we can follow the biological side of the human and animal organisations by looking at their sensory life.

The sense life unfolds — as I have indicated years ago how it is partially a fact of outer science — not only in the usually claimed five senses, but in a clear discernible number of twelve human senses. Now, we are only talking about human beings. For those who want to understand if one could speak about twelve senses in the same way as for five or six — from seeing, hearing smelling, tasting, feeling or touch — for those it is valid that one can speak for instance about the sense of balance, which we recognise inwardly whether we stand on two feet or only one, whether we move our arms in one way or the other, and so on. By our position as humans in the world, we are in equilibrium. This equilibrium we accept, although from a much duller position than that of our perceiving through the senses about what takes place in the process of seeing; so that we may speak about the sense of equilibrium as we speak about a sense of seeing. Let us be clear about this. When we speak about the sense of equilibrium we turn ourselves more towards our own organisation, we perceive inwardly, while with our eyes we look outwardly. However, this experience of equilibrium fosters its basis as a sense perceptible function. In the same way we can expand the number of senses on the other side. When we merely hear something, the function of the human organism is really different to when we, as it were, hear directly through the ear, and then explore what becomes indirectly perceptible to us, through speech. When we follow the words of others with inner understanding it doesn't involve mere judgement, but a process of judgement comes out of a perceptive process, a sense process; so we need to speak about it as having a sense of speech — or a sense of language, a sense of the word — just as we have a sense of hearing. In other words, we must, if we consider the words more anatomically-physiologically, presuppose there is within the human organization a special (sensory) organization which corresponds to the hearing of what had been spoken as well as hearing inarticulate sounds. We must assume a special organisation for the sense of speech, which is quite similar to a sensory organisation, for example the organisation of sight or of hearing.

We may also, when we go to work without prejudice, not say: we get to know that a person is standing in front of us, when we see there is something in the external space shaped like a nose, like two eyes and so on, and through an analogy conclude that a person is stuck in there, because we see that in us there is also a person, revealed outwardly through a nose, eyes and so on. Such an unconscious conclusion in reality doesn't form the basis; it is rather the direct entry into others which corresponds to something special within them which can be compared to a sensory organisation, so that we can speak about a sense of Self (Ichsinn). When we look through the functioning of people in this unprejudiced way, then we need to, with the same authorization with which we spoke about the sense of hearing, of taste and so on, speak about the organisation of perception for words, about an organisation of perception for thoughts, for an organisation of the Self — not for one's own self, because for one own Self it is dependent on something quite different. Further, we must speak about a sense of movement, because it is something quite different, whether we call it movement or rest. Likewise, we must speak about a sense of life — ordinary science already speaks partly about that.

When we determine the number of sense organisations, we arrive at twelve human senses. Of these, several are inner senses, because we involve the inner organism — how we feel and experience the sense or equilibrium, sense of movement and so on — while observing it. However, qualitatively the experience with observation of the inner organization remains the same for the seeing, hearing or taste processes. It is important to see things only in the right context.

If the starting point is from a human angle of a complete physiology of the senses then certain biological phenomena from the human realm, on the one side, and the animal realm on the other, reveal a particular meaning. This meaning can exist even if you admit to everything which has been presented by recent research, even from Haeckel, regarding the morphological and also physiological human organisation in relationship to that of the animal. Here the most impossible misunderstandings come about. It is believed, for instance, that Anthroposophy must oppose Haeckel, simply on the grounds that it rises from mere observation through the senses to the empirical observations of the spiritual; it is believed that Anthroposophy must from this basis change Haeckelism. No! — What needs to be changed in Haeckelism must be changed out of natural scientific methodology, so Anthroposophy doesn't need to argue here because one can have discussions, as scientific researchers, with Haeckel.

What Anthroposophy has to offer refers to quite other areas. It is correct to emphasize that by counting the bones of the higher animals, there's no differentiation to the number in humans. The same goes for the muscles. This gives no differentiation between human and animal organisations. If, however, we proceed biologically, we discover real differentiation. We find that we can attribute a special value to the essential way the human organism is placed differently in the cosmos to the animals. When we observe the higher animals, we have to admit that their essential aspect is that the axis of their backbone is parallel to the earth's surface while by contrast, humans in the course of their life make their horizontal spinal axis vertical, which means an important function in the life of humans are to stand upright. — I know objections can be raised: there are some animals which have more or less of a vertical spinal axis. This is not the salient point as to how it is excluded through outer morphology, but how the entire organisation is adjusted. We will also see how with certain animal, bird types or even mammals, where the spine can be brought into more or less of a vertical position, a kind of degenerative effect appears in their total predisposition, while with people there is already a predisposition for the spine to be vertical.

When I mentioned this some years ago in a lecture in Munich, a man educated in natural science approached me, who I could naturally understand quite well, and said: 'When we sleep, we do have a horizontal spine.' — This does not matter, I replied, but what matters is in the relationship to the situation, let's say, of the bone in the leg or foot to the rest of the bodily structure and to the whole cosmic relationship of mankind, and how this is processed by the human or by the animal. Indeed, the human's spine is horizontal during sleep, but this position is outward; inwardly the human is so dynamically organized that he can bring himself into a condition of equilibrium where the spine is vertical. When animals come to such a state of equilibrium, they are degenerating in a certain sense, or they tend towards developing some human-like functions and as a result prove, what I want to present now.

We can say that by the human being purely functionally, out of the total dynamic of his being in the course of his first year of life, forming his spinal axis vertically, he has brought himself into another relationship of equilibrium cosmically than the animal. However, every being is created out of the cosmic totality, and one can say, and adapt — I don't want to enter further into this.

When we track the formation of single bones, for example the ribs or head bones further, then we also gain the morphological possibility of how the formation of ribs or head bones in a man or dog adapt to the vertical or horizontal spine. Because the human being finds himself in a vertical position he lives in contrast to the animal, who stands on four legs, in quite a different state of equilibrium, in quite a different cosmic relationship.

Now we must try to clarify the problem from the other side, what actually happens in the sense's processes in a person and what happens to him with reference to the sensory process. Due to our limited time I wish to speak only with indications, but this can also be translated into a completely precise biological-physiological terminology.

Let's take the process of seeing. We could create divisions into what the specific function of the organ of sight is, and into what happens further as a continuation within the physical;

one could call it an analogy of the optical nerve of the eye which loses itself in the inner nervous system. Thus, we could differentiate, on the one side, the process of sight itself, and then everything connected to it in the totality of experience. In the direct present process of vision there is also the imminence of the image perceptibility; when we look at something then we don't separate the imagery from the vision. Turning our eyes away from what we looked at, we then retain a kind or imaginative remnant which clearly shows a relationship with the vision's observation. Whoever can really analyse this will see the differences between the imagined remnant obtained through the organ of vision as opposed to what happens with the hearing process. We have within us an experience of the process of sight, one could call it, in a dualistic way. First being more turned to what actually is observed through the senses and then turning again to the remnant within imagination which remains more or less as a manifested memory.

Let's take everything that lives in the inner image perceptibility of human beings, which depends on the five senses. The one which is the most dependable is of course the process of sight. Only a ninth of what is found through vision, is found through the hearing process. When we consider soul life, there is even less found in it than the seeing and hearing processes, and so on. We know, that in addition to the image perceptibility which leads to lasting memory, it also plays a role but in reality, less than with the seeing and hearing processes.

Now we can pose the question: Is there also for the more hidden senses, like the sense of equilibrium or sense of movement, this duality which is found in the observation perceptibility and image perceptibility? With a truly unbiased physiology and psychology the duality is there for instance in the sense of equilibrium, but the connection is seldom noticed. In the lectures I've just given, I spoke about the mathematical geometric relation of finding oneself upright in relationship to space. We construct relationships to space. What is it actually, that we are doing here? It is connected to the entire person just as it is when with the process of sight, the observed element is clearly separated from the image perceptibility, because we keep the imagination inside. We don't observe the colour outwardly, but we experience the qualitative aspect of the colour, of colour tones, and what lives as a feeling, a feeling I have towards warm or cold colours strongly within myself.

We can now say the following: 'I want to instantly see a show of all the soul images which I've acquired through my life, which I can see through my eyes.' We would then enter into a visual system of the soul. We would, without having outer sight, rise inwardly into a kind of reconstruction of the visual process. If you apply this same kind of consideration to the sense of equilibrium, you obtain everything which you have experienced through the sense of equilibrium in your own organism, rising within, which corresponds to the geometry in outer life. [*In the next few sentences some word-gaps followed in the stenographic text. The omissions were filled in with the help of the lectures of Dr Steiner on the 16th March 1921, 29 September, 1 and 3 October 1920 in GA 322*] Mathematic and mechanistic laws have not been discovered by outer experience. Mathematic and mechanistic laws have been acquired though inward construction. If you want to recall mechanistic laws you have to access them through the image perceptibility of your sense of equilibrium. The entire human being becomes a sense organ and thus inwardly creates the other pole of what had been observed. For instance, we create mathematics and we believe we have a purely a-priori science. However, mathematics is no pure a-priori science. We don't notice that what we are experiencing as a sense of balance is what we translate into mathematical geometric imagination, like the observation through sight is translated into the imagination of the observed sight. Without noticing the bridge, we have created the sense of balance through maths or through mechanics.

When you think about this, you would understand the innermost relationship of the human organism with its position of equilibrium in the cosmos. Then you could say to yourself: With the animal, which stands on its four legs and which has been given its equilibrium position and sense of equilibrium, the animal must experience equilibrium in quite a different way within itself than a person does mathematically. We find the mathematical simply as a result of us being placed in the cosmos.

We talk about three dimensions because we are positioned in three dimensions in the cosmos. However, the vertical dimension we have only achieved in the course of life. We have placed ourselves into the vertical position. What we have experienced in our earliest childhood reflects later in us as mathematics; it only doesn't develop as quickly as the seeing process. The reflection of the experience of equilibrium goes on in the course of life. In childhood we have a very strong experience of the sense of equilibrium, when we go over from crawling to walking and standing. This reflects in later life and becomes visible as mathematics and mechanics. We often take mathematics as something woven out of ourselves. This is not so. It comes out of the observation of our organism. Why then are there certain thoughts which can be related to the cosmos, which then can create an entire construction in thought, a 'thought-edifice?' That is merely a result of human beings standing within the cosmos. When we now compare the position of equilibrium between the animal, in his relation to the cosmos, with that of the human being, then we could say: We have with the animal the bondage with the earth organisation and we have with the human being the uprightness, being 'lifted-out' of the earth organisation. What we express as individualized thoughts result from our human organisation having an individual position of equilibrium.

Thus, this actual act of placing oneself in the cosmos is not something which emerges from the organism itself as is found with the animals, but it is something formed within the human organism which is only achieved in the course of the first seven years of life and goes right into the organs. As a result, we have this polarity between animals and humans that on the one hand humans stand upright and walk vertically. This is quite a cosmic position which lives in the human being, to which everything now has to adapt, and which distinguishes it from the animal. On the other hand, thoughts appear in the soul, thoughts which go beyond the sensual perception, beyond what is sensed with the five senses, extricating themselves from that. Just as the human being frees itself in its cosmic position from the earth, so the human thoughts extricate themselves from the bondage of the sense world, they become free in a certain way.

We must — for anthroposophy it is a definite but here I want to present it more as a hypothesis — we must see in the human being, through his upright spinal axis, a certain position of equilibrium which separates him from animals, and on the other hand we must regard thoughts of a particular imaginative form, as specifically human. Whoever examines such things from an Anthroposophic viewpoint — it could still become more or less relevant — will see how the human being's particular development of his sense of equilibrium and sense of movement achieve more towards a free system of thought than the case is through the eyes and ears, and we also gain insight of the human being requiring an inner organisation for this. The human being simply has an organisation within itself which is not yet found in the animal — this can also be proven materially — which simply supports ideas which are torn free from the earth's bondage to which the animal is bound, but which is limited by the special equilibrium position in humans. Therefore we can say: by the human standing upright, he has created an organ for abstract thinking.

So we have with the upright related organisation of human organs a different situation to those in the animals, which have organs too. Through the upright position the nerves and blood organisation work in a different way under the influence of the equilibrium position so that in the human something appears which can't be brought about in the animal. We discover this relationship really in the physical organization of humans and not as mere dynamism. This is of fundamental importance. Just imagine what happens in the evolution of an organization as a result of a change in their position of equilibrium, how it appears in the animal, how it is in the human being, what changes there for instance with reference to the upper and lower leg, hands and so on. Just imagine what it means that the human being is a two-hander and not a four-footer. The human being is fitted out with the same forms as the animal but they are in different positions and as a result are changed, metamorphosed forms. This can also simply be anatomically demonstrated if the necessary tools and experimental methods qualify. We are looking for such tools and experimental methods in our institute in Stuttgart. In any case before these methods can be empirically found externally, the differences need to be arrived at firstly through imaginative observation. Therefore, Anthroposophy is not useless with reference to research into the finer areas of the human, animal and plant forms, because science can't discover these things through imagination. Once they are discovered they can be verified by science.

When one looks at how another position of equilibrium reorganises the organs, one also finds that certain organs are changed in such a way that they become the human organ of speech and the organism becomes capable of creating speech.

Through this you have gained an insight into the extraordinary organisation of mankind, which has simply come about because of it being an upright being, having repercussions right into matter. Also, in relation to the physiological organ of speech it has contributed — where an outer morphological distinction between man and animal can be determined — which after all shows the difference between man and animal biologically.

Here you have a few suggestions which can indicate the way how, in an outer lay method, research may be done which can also be actually researched scientifically. What I wanted to bring I could only briefly sketch here. However, continue to think about these ideas and the results will actually be a way for science to research the differentiations between the animal and human organisations in a biological relationship.

"ANTHROPOSOPHY AND PHILOSOPHY."

Lecture Three by Rudolf Steiner given in Berlin, 7 March 1922

M y dear venerated friends! It is always difficult when you have a serious scientific conscience to translate the traditional expression of "Logos" into some or other younger language. We usually employ "Word" to translate "Logos" as is commonly found in the Bible. However, when we have the word "Logic" in a sentence we don't use "Word" but rather think about "Thought," as it operates in the human individual and its laws. Yet when we speak about "philology" we are aware that we are developing a science which is derived from words. I would like to say: what we have today in the word "Logos" is basically in everything which is philosophic. When we speak about "philosophy", we can, even though defined as experience in relation to the Logos, sense how a reflection of these undetermined experiences are contained in all that we feel in "philosophy".

Philosophy implies that the words — which no doubt came into question when philosophy was created, that only words were implied — indicate a certain inner personal experience; the word philosophy points to a connection of the Logos to "Sophia"; one could call it a particular, if not personal, general interest. The word philosophy is less directly referred to as possessing a scientific nature but rather an inner relationship to the wisdom filled scientific content. Because our feeling regarding philosophy is not as sure as in those cases when philosophy, on the one hand was included with, I'd rather not call it science, but scientific aims, and on the other hand with something which points to inner human relationships; so we have today an extraordinarily undefined experience when we speak about philosophy or involve ourselves with philosophy. This vague experience is extremely difficult to lift out of the depth of our consciousness if we try to do it through mere dialectical or external definitions, without trying to enter into the personal experience which ran its course in the consequential development. To such an examination the present will produce something special.

If we look back a few decades at people in central Europe, the involvement they were looking for with philosophy was quite a different experience, in central Europe, as it is today in the second decade of the twentieth century, where we basically have lived through so much, not only externally in the physical but also spiritually — one can quietly declare this — than what had been experienced for centuries. When one looks back over the experiences, of — if I may use a pedantic and philistine expression — the philosophic zealot of the fifties, sixties and seventies of the nineteenth century, perhaps even later, which the central Europeans could have, it is essentially as follows. Looking at the time of German philosophy's blossoming, you look back at the great philosophic era of the Fichtes, Schellings, and Hegels; surrounding you there had been a world of the educated and the scholars, a world which this philosophic era thoroughly dismisses and which in the rising scientific world view sees what should be taking the place of the earlier philosophic observations. One admires the magnitude of the elevation of thoughts found in a <u>Schelling</u>, one admires the energy and force of Fichte's development of thoughts, one can perhaps also develop a feeling for the pure comprehensive, insightful thoughts of <u>Hegel</u>, but one would more or less consider this classical time of German philosophy as something subdued.

Besides this is the endeavour to develop something out of science which should present a general world view, right from the striving of the "power/force and matter/substance people," to those who carefully strive to find a philosophic world view out of natural scientific concepts, but who lean towards the former idealistic philosophy. There were all kinds of thoughts and research in this area.

A third kind of thinker appeared in this sphere, who couldn't go along with the purely scientifically based world view but could on the other hand also not dive into solid thought of the Hegel type. For them a big question came about: How can a person create something within his thoughts, which originate in himself, and place this in an objective relationship to the outer world? — There were epistemologists of different nuances who agreed with the call "Back to Kant", but this way to Kant was aimed in the most varied ways; there were sharp-witted thinkers like Liebman, Volkelt and so on, who basically remained within the epistemological and didn't get to the question: How could someone take the content of his thoughts and imaginative nature from within himself and find a bridge to a trans-subjective reality existing outside human reality?

What I'm sketching for you now as a situation in which the philosophic zealots found themselves in the last third of the nineteenth Century, which didn't lead to any kind of solution. This was to a certain extent in the middle of some or other drama during a time-consuming work of art, to which no finality had been found. These efforts more or less petered out into nothing definite. The efforts ran into a large number of questions and overall, basically failed to acquire the courage to develop a striving for solutions regarding these questions.

Today the situation in the entire world of philosophy is such that one can't sketch it in the same way as I've done for the situation in the last third of the nineteenth century, in its effort to determine reality. Today philosophic viewpoints have appeared which, I might say, have risen out of quite different foundations, and which make it possible for us to characterise it in quite a different way. Today, if we wish to characterise the philosophic situation, our glance which we have homed in the second half of the twentieth century comes clearly before our soul eyes, namely such sharply differentiated philosophic viewpoints of the West, of central Europe and Eastern Europe. Today things appear in quite a different way which not long ago flowed through our experience of the philosophical approach to be found in three names: Herbert Spencer — Hegel — Vladimir Soloviev. By placing these three personalities in front of us we have the representatives who can epitomise our philosophic character of today. Inwardly this had to some extent already been the case for some time, but these characteristics of the philosophic situation only appear today before the eyes of our souls.

Let's look at the West: Herbert Spencer. If I want to be thorough I would have to give an outline of the entire course of philosophic development, how it went from Bacon, Locke over Mill to Spencer, but this can't be my task today. In Herbert Spencer we meet a personality who wanted to base his philosophy on a pure system of concepts, as is determined in natural science. We find in Spencer a personality who totally agrees with science and out of this agreement arrives at a conclusion: 'This is the way in which all philosophic thought in the world must be won by natural science.' So we see how Spencer searched in science to determine certain steps to understand concepts, like for example how matter is constantly contracting and expanding, differentiating and consolidating. He saw this for instance in plants, how the leaves spread out and how they drew together in the seed, and he tried to translate such concepts into clear scientific forms with which to create his world view. He even tried to think about the human community, the social organism, only in such a way in which his thoughts would be analogous to the natural organism. Here he suddenly became cornered.

The natural human organism is connected to the confluence of everything relating to it from the surrounding world, through observations, through imagination and so on. Every single organism is bound to what it can develop under the influence of the nervous or sensory system (sensorium). In the human community organism Herbert Spencer couldn't find a sensorium, no kind of centralised nervous system. For this reason, he constructed a kind of community organism, totally based on science, as the crown of his philosophic structure.

What lay ahead for the West with this? It meant that scientific thought could reach its fully entitled, one-sided development. What lay ahead was the finest observational results and experimental talents developing out of folk talents. What came out of it was interest created to observe the world in its outer sensory reality into the smallest detail, without becoming impatient and wanting to rise out of it to some encompassing concepts. What came out of it was also a tendency to remain within this outer sense-world of facts. There was what I could call, a kind of fear of rising up to one encompassing amalgamation. Because they could do nothing else but exist in what the sense world presented to them, simply being pushed directly into the senses here in the West, there appeared the belief that the entire spiritual world should be handed over to the singular faiths of individuals, and that these beliefs should develop free from all scientific influences. Religious content was not to be touched by scientific exploration. So we see with Herbert Spencer, who in his way took up the scientific way of thought consequentially right into sociology, earnestly separated, on the one hand, from science, which would proceed scientifically, and on the other hand with a spiritual content for people who wanted nothing to do with science.

Let's go now from Herbert Spencer to what we meet with Hegel. It doesn't matter that Hegel, who belonged to the first third of the 19th Century, was outwitted during the second third for central European philosophy because what was characteristic for Middle Europe was most meaningful in what exactly had appeared in Hegel. Let's look at Hegel. Already in his, I could call it, emotional predisposition, lies a certain antipathy against this universalist natural scientific way with which to shape the world view as Herbert Spencer had done in the West, but of course had been prepared by predecessors, both by scientific researchers and philosophers as well. We see how Hegel could not stand Newton and was unsympathetic to his unique way by thinking of the world-all as totally mechanical, how he rejected Newton not merely in terms of the colour theory but also in his interpretation of the cosmos. Hegel took the trouble to go back to Kepler's planetary movement formulations, he analysed Kepler's formulations about planetary movements and found out for himself, that Newton had actually not added anything new because Kepler's formulation already contained the laws of gravitation. This he applied from the basis of a scientifically formulated thought, while with Kepler it had resulted more out of a spiritual experience, which he saw as encompassing and that one could try to grasp the outer natural scientific through the spirit. Kepler is for Hegel simply the personality who is capable of penetrating thoughts with the spirit and building a bridge between what is acceptable scientifically, and what simply has to be believed according to the West, and which is also capable of lifting science into the area which for the West is limited to belief.

From this basis Hegel, in tune with Goethe, strongly opposed the Newtonian colour theory. We can see how the Hegelian system had a kind of antipathy against what appeared quite natural in the Newtonian system. For this Hegel had a decisive talent — to live completely in a thought itself. For Hegel, Goethe's utterance to <u>Schiller</u> was obvious: "I see my ideas with my eyes." It appears naive, however, such naivety, when considered correctly, comes out of the deepest philosophic wisdom. Hegel would simply not have understood how one could state that the idea of the triangle is not to be grasped, because Hegel's life went completely — if I might use the expression — according to the plan of thinking. For him there was also a higher world of revelations, a world of higher spirituality, which gradually casts its shadow images on a plane which is filled with thoughts. From up above the spiritual worlds throw their shadow images on the plane of the human soul, on which human thought can develop. Through this the idea of higher spirituality came about for Hegel, that on the plane of the soul it is shadowed as thoughts. Hegel was inclined to experience these thoughts as fully spiritual, and he also experienced natural events not in their elementary present time, but he saw them in mental pictures, thrown on to the plane of the soul.

So it is impossible in Hegel's philosophy to separate, in an outer way, wisdom from belief, which was quite natural in the West. For Hegel his life task was the unification of the spiritual world (which the West wanted to simply refer to as part of the large sphere of belief) with the sensory physical world, into such a world about which one can have knowledge. This means there is no longer knowledge on the one hand and belief on the other; here the human soul faces the great, meaningful problem: How does one find during earthly life the bridge between belief and knowledge, between spirit and nature? To a certain extent it was the tragedy of Hegel that the problem he posed in such a grandiose manner, he wanted to understand actually only on the level of thinking, that he wanted to understand the experience of the inner power, the inner liveliness of thinking, but he could not grasp anything living from the content of thought.

Consider Hegel's logic — he wanted to return repeatedly to the concept of the Logos! He felt that when we actually wanted to attain a true understanding of the Logos, then the Logos must be something which is not merely something thought, but a real activity which floods and works through the world. For him the Logos did not only have an abstract, logical content, but for him it became real world content. If we look at one of the three parts of his philosophy, namely his "logic" we only find abstract concepts! So it is terribly moving for someone who enters on the one side into the Hegelian philosophy, with his whole being, and has the fundamental experience: that which can be grasped through the Logos, must be penetrated with the creative principle of the world. The Logos must be "God before the creation of the world" — to use an expression of Hegel.

This is on the one side. Now how did Hegel develop this idea of the Logos on the other side? He starts with "being" and arrives at "nothing", goes from "becoming" to "existence." He arrives at the goal through the causality, to the belief that certain phenomena are best explained in terms of purpose rather than cause. One can look at the all the concepts of Hegel's logic and ask oneself: Is that what, "before the beginning of creation as the content of the divine" could have been there? This is abstract logic, the demand of the creative, the logos as postulate, but as a purely human thought postulate! One finds this tragic. This tragedy goes further, for the Hegelian philosophy is deemed as valid. Yet it contains instances where through action new life can germinate. It contains sprouts. Hegel saw his redemption in this: being — nothing — becoming — existence.

When people are presented with Hegel, they say: 'This is a dark one, we don't need to be lured into it.' However, when one makes the effort to allow one's inner soul to enter into it, to experience the concept inwardly, as Hegel tried to experience it, then all the ideas of empiricism and rationalism disappears, then thought experiences and the one who is thinking is directly thought of. Whoever goes along with it finds the impetus of loosening the thoughts from the abstraction and take Hegel's logic as the sprouts which can become something quite different, when they become alive. For me Hegel's logic looks like the seed of a plant in which one can hardly see what it will become and yet still carries the most varied structures possible within it.

For me it appears that when this seed sprouts, when one lovingly cares for it and plants it into the soul's earth through anthroposophical research, then what emerges is that thought can not only be thought but can be experienced as reality. Here we have the central European aspect.

If we now go to the East, we have in Vladimir Soloviev a man who is able like no other philosopher, to become gradually more the content of our own philosophic striving, who must now become so important to us because we allow the particularities of his character to work in on us. We see in Soloviev both the European-eastern way of thinking, which is of course not Oriental-Asiatic. Soloviev absorbed everything which was European, he only developed it in an Eastern fashion. What do we see being developed in terms of human scientific striving? Here we see how actually this method of thinking, found mostly in the West by Herbert Spencer, which Soloviev basically looked down on, is something against which the truth and knowledge he was seeking, could so to speak be illustrated. In comparison, what he actually presents is a full experience of spirituality itself. It appeared in full consciousness to him, it appeared more atavistically, subconsciously, yet it is an experience in spirituality itself. It was more or less a dreamlike attempt to knowingly experience what in the West — here quite consciously — was transposed into the realm of belief. So we find in the East a discussion which can be experienced in an imprecise way, which looks like a one-sided experience which Hegel wanted to use to cross the bridge out of the natural existence to the spiritual world.

If a person delves into the spiritual development of someone from central Europe, like Soloviev, then he will primarily have an extraordinary uncomfortable feeling. He is reminded of an experience of something misty, mystical; an overheated element in the soul life which doesn't arrive at concepts, which can externally leave him empty completely, but which can only be experienced inwardly. He senses the entirely vague mystical experience, but he also finds that Soloviev makes use of conceptual forms and means of expression which we know, from Hegel, Humes, Mills, even those of Spencer, but only as illustrations. Throughout one can say he doesn't remain stuck in the mist but through the way with which he treats religious aspects as scientific, how he searches for it everywhere and unfolds it as philosophy, he can evermore be measured and criticised according to the philosophic conceptual development of the West.

So we find ourselves today in the following situation. In the West comes the striving to formulate a world view scientifically; science is on the one side and the spiritual on the other side and wrestle in the centre with the problem of how to create a bridge to include both, to express it imprecisely, as Hegel said: "Nature is Spirit in its dissimilarity," "Spirit is the concept of when it has returned again to itself." In all these stuttering expressions lie the tragedy that Hegel could only care for abstract ideas, which he strived for. Then in the East, with Soloviev we see how it was somewhat still maintained, how well the church fathers wanted to save it in terms of philosophy, before the Council of Nicaea. It places us completely back in the first three post Christian Centuries of the West.

So we have in the East an experience of the spiritual world, which is not able to soar up into self-owned terminological formulations, formulations and concepts used by the West in which they express themselves, and as a result remain in vague, somewhat extraneous, foreign expressions.

So we see how the threefold nature of the philosophic world view unfolded. By our tracing how the threefold philosophic world view was formed through the characteristics and abilities of humanity in the West, the centre and the East, we can see that we are obliged today — because science as something embracing must spread over all of mankind — to find something which can lift it above these various philosophic aspects which basically still provide elements where philosophy is still a human-personal matter. We see today in different ways in the West, central Europe and the East, how they love wisdom. We understand that in ancient times, philosophy could still be an inner condition of the soul. Now however, in recent times, where people are strongly differentiated, this way of loving wisdom expresses itself in a magnitude of ways. Perhaps we could realise due to this, what we have to do ourselves, particularly what we have to do in Central Europe, where the most tragic and intensified problem is raised even if it is not regarded in the same way by all philosophic minds.

If I want to summarise all of what I have brought into a picture, I would like to express it as follows. Regarded philosophically Soloviev speaks like the old priest who lived in higher worlds and who had developed a kind of inner ability to live in these higher worlds: priestly speech translated philosophically is what one encounters all the time with Soloviev. In the West, with Herbert Spencer, speaks the man of the world who wants to enter practical life — as it has come out of Darwinian theory — to expand science in such a way that it becomes the practical basis of life. In the Middle we have neither the man of the world not the priest: <u>Fichte</u>, Schelling, Hegel have no priestly ways like a Soloviev. In the Middle we have the teacher, the educators of the people and it is also here where the German philosophy emerged, for example, from religious deepening; because the priest became the teacher once again. The educated also adheres to the Hegelian philosophy.

We see recently — as with Oswald Külpe — how it has happened that philosophy, when it was already lost, is no more than a summary of the individual sciences. From inorganic science you can ask — what are the concepts? From organic science you can ask — what are the concepts? Likewise with history, with the science of religion, and so on. One collects these concepts and forms a separate abstract unit. I would like to say that the subject of the teaching in the separate sciences should create the totality of teaching. This is what the science in the Middle must basically come to after the entire assessment.

If we look back at what has happened, we see with Herbert Spencer the unconditional belief in science, the belief for the necessity to cling to observation, experiment and a thinking mind, which can be experienced through the observation and experiment; and one is mistaken about the contradiction which appears here, when the acquired concepts can be applied to the social organism and — although these do not have the most important characteristics of a natural organism, the sensorium — they are nevertheless grasped with the same concepts which arise in natural existence. We see the inclination to the natural sciences so strong that some characters — like Newton — became one-sidedly stuck to the mechanistic and even satisfied their soul-striving with it. It is generally known that Newton had tried in a one-sided mystical way to clarify the Apocalypse; besides his scientific world view he had his own mystical needs.

Let's look, for example, at everything which has arisen from natural science and what it gradually in the course of the 19th Century has subconsciously taken over in Central Europe; because in Central Europe science has simply followed the pattern of the Western scientific way of thinking. There is a tendency not to take notice of it, but still all points of view are modelled on the Western pattern. How wild the people become when someone tries to apply Goethe's way of thinking in physics in contrast to them taking shelter under Newton!

How does the development happen in biology? Goethe created an organism for which the integration into its concepts depended on an understanding of a mathematical nature. Time was short to obtain a biology more appropriate to modern thinking than to that of olden times. The progress in the 19th Century in central Europe however brought about not the Goethean biology but Darwinism, which was interspersed with concepts contrary to those of Goethe, like the concepts of the 16th Century opposed to those of the 18th Century. Only in Central Europe did these concepts develop; in the West people remained with those concepts that sufficed for the understanding of nature. So it happened that certain concepts in the West simply were not available and simply got lost because people in Central Europe had adopted western thinking. For example, that a thought, a lively thought, can form a concept of grasping a reality, quite apart from empiricism, as it had happened with Hegel — this is not present in Central Europe; it got lost because the central European thinking was flooded by western thinking.

So we have the task in Central Europe to look at what scientific thinking can be. Anthroposophists resent it when this scientific way of thinking is cared for with as much love as for the researcher himself. Nothing, absolutely nothing will be said by me in opposition to scientific thinking; if someone believes this then it is a misunderstanding. However, I must understand the scientific way of thinking in its purity and then also try to characterise it in its purity. Now these things are presented to those who confront scientific thinking with impartiality — somewhat like a western researcher will present them, like <u>Haeckel</u> in his genial way did it — these results are presented in a western way of research, when they are thus left and not reinterpreted philosophically, not given as solutions, not as answers, but are presented above all as questions. The totality of natural science does not gradually become an answer to a question for the impartial person, because it turns into the great world question itself. This is experienced everywhere: what is now being researched in the most beautiful way by these researchers — for my sake right up to atomic theory, which I don't negate but only want to put it in its correct place — this comes to a question and out of the West a great question is posed to us. Where does this question come from?

When we link our gaze to the outer world and only turn to the observation of the given elements, we don't fathom its complete reality. We are born as human beings in the world, are constituted as such, as we already were before and take part in the reality by looking at ourselves in our own inner being. As we look then at the outer world, the sense perceptible objects — we find that part which is living in us, is missing in reality, as we can only through human struggle connect to the other half-reality, which observes us from the outside. If we look towards the West, so we see the half-reality is researched with particular devotion; however, it only provides a number of questions because it's only a half-reality. So on the one side there appears only one half of reality as a given; if one really looks at it, it raises questions. In Central Europe you discover examples of questions which Western thinking can answer, and one tries to push through to thinking. That is the Hegelian philosophy.

In the East one felt that which lives above the thought, which works down into the thought; but one couldn't come as far as awakening it to life, that so to speak the flesh could also sustain a skeleton. Soloviev was able to develop it in flesh, muscles and even blood in his philosophy — but the skeleton was missing. As a result, he took Hegel's concepts, those of Humes and others, and built in a foreign skeletal system. Only when one is in the position of not using a foreign skeletal system then something comes about which can be lived through spiritually. So, however, as it happened with Soloviev, it leads to a shadowed existence because it didn't manifest into a skeletal system which could as a result be descriptive. If one doesn't want to remain with building only an outer skeletal system but live spiritually and prepare oneself through strong spiritual work, then one develops for oneself an inner skeleton within spiritual experiences; one develops the necessary concepts. For this, various exercises have been given in my writings, "Occult Science" and "Knowledge of the Higher Worlds" and in others. Here one develops what really can become a conceptual organism. This is then the other side of reality, and this side of reality has its seed in the eastern philosophy of Soloviev.

In central Europe there is always the big problem of striking a bridge between nature and the spiritual. For us it has at the same time become a meaningful historical problem: to strike the bridge between West and East, and this task must stand before us in philosophy. This task also directs itself into Anthroposophy. If Anthroposophy becomes capable of inward thought experiences developing into living form, then it may on the other side experience quite materialistic natural phenomena as they are experienced in the West, because then it will not be through abstract concepts but through living scientific circles that the bridge is built between mere belief and knowledge, between knowing and subjective certainty. Then out of philosophy a real Anthroposophy will develop, and philosophy can be fructified from both sides by these living sciences. Only then would Hegel's philosophy be awakened to life, when through the anthroposophical experience you let the blood of life be spiritually added to it. Then there won't be a logical base which is so abstract that it can't be "Spirit on the other side of Nature", as Hegel wanted it, but that it really can be grasped, not as abstraction but as the living spirituality of philosophy.

This gives Anthroposophy the following task. How must we, according to our present viewpoints, which lie decades behind Hegel, strike the bridge between what we call truth on the one side, which must encompass all of reality, and that which we call science on the other side, which also must encompass the entirety of reality? Briefly, the problem must be raised — and that is the most important philosophic problem in Anthroposophy: what is the relationship between truth and science?

This is the problem I wanted to present in the introduction today at the start of our consideration, which I believe you will now understand.

"ANTHROPOSOPHY AND PEDAGOGY."

Lecture Four by Rudolf
Steiner given in Berlin,
8 March 1922

My dear venerated guests! To render the Anthroposophical world view understandable, it is always accused of having its ideas and results as based on research by people who first need to be schooled for it, therefore results of Anthroposophical research can't be verified from the outset by anyone and that nevertheless these observations are presented to unprepared people.

Yet this accusation, however justified it seems, belongs to the most unjustified ones that can be made against the Anthroposophical Movement, because it doesn't stipulate that every single person is to be immediately directed towards becoming a researcher in a supersensible area, but it deals with their research results being presented in such a way that every other individual can prove it for themselves, simply with ordinary human understanding and ordinary healthy logic. In any case this doesn't make it unnecessary that at least the first steps towards supersensible research must be striven for, and therefore there are indications in various publications which have been mentioned here already. Everyone can to a certain degree become an anthroposophical researcher — simply out of the conditions of present civilisation — but as proof of results of anthroposophical research this is not necessary because this proof can simply come through a healthy human understanding. One of these areas in which results can really be practically proven, is in the pedagogical area.

Dear friends! The Anthroposophical world view for a long time had to work purely through people coming continuously closer to ideas about the supersensible, before it became possible to bring them into present-day cultural conditions, into practical life, where they felt themselves particularly ready to actually penetrate. This was only possible in a limited area — and also only to a small degree — when Emil Molt founded the Waldorf School in Stuttgart, whose office was given to me. Already before that, as shown in the small publication "The education of the child from the point of view of spiritual science", the attempt was undertaken to represent certain educational principles from the basis of Anthroposophy. Only through the founding of the Waldorf School did it become possible to apply these things in practice, and since this time it is also possible to carry out the pedagogical-didactic side of Anthroposophy in detail. It will of course not be possible for me to give more than a few indications in this introductory lecture, but I think other lectures during these days would be able to accomplish more.

Whatever is taken up through anthroposophic ideas, when they are simply verified through healthy human understanding, is not merely theoretical observation, are no mere ideas of abstraction which one can have in order to satisfy some or other need for theoretical knowledge. No, these conveyed ideas are being created out of an anthroposophical source, this is real human power; this is something which passes into the whole person, this makes love more intense, transforms human vigour. While the ideas and thoughts of usual science, which only draw on the sense world, have their peculiarity by being in the service of theoretical interests, and being of the sense-world its characteristic is to only relate practical interests to the sense world, by contrast it is characteristic of the ideas from anthroposophic research that their results work on the entire human being, on his empowerment, on his — if I might call it so — life skill, on his understanding of life, and it is this understanding which enables him to grasp the most varied of life's opportunities. If one takes on this life and fructify it through anthroposophical ideas, one can see how the actions of people, when they allow themselves to be directed by these ideas, acquire greater power, greater urgency and so on. This is what must be especially treasured in the pedagogic didactic sphere.

When we founded the Waldorf School we didn't have the opportunity to choose the outer conditions for the education and teaching of our children. In the present it is repeatedly asserted that for a satisfactory education, satisfactory teaching to be established, then some or other place for the school, for the educational institute or its equivalent, must be chosen. Certainly, much is said about these claims and in practice they are to some extent successful. We didn't have all of this. The next thing was the attempt to use the given circumstances and start with the children in the Waldorf-Astoria Cigarette factory. Next, we had a very specific kind of material environment for the children, we had to house ourselves in a place which was obviously hardly suited — it had previously been a tavern — to begin our teaching and education. So we couldn't rely on anything but on what began on a purely spiritual basis for the pedagogical and didactic aspects themselves.

Here it must be stressed again: while Anthroposophy doesn't strive for an abstract head knowledge — if I might use this expression — but an insight into the world and its secrets, it involves the whole human being, possibly leading to self-knowledge, to a self-understanding which one can't achieve in some or another theoretical sphere. In the end all education, all teaching is based on the understanding of the human being, which is proven in the relationship of the teacher, the educators to the emerging, growing adolescent, to the child. For this reason, our Waldorf pedagogy is developed upon an intimate knowledge of the growing person, the child. I only need to give one detail in which it can be seen how true insight into the whole human being must prove in practice.

Today we have a psychology which has more or less been proven by recognized science. However, this psychology theorizes around many questions which always leave unsatisfactory results. They pose the following question, for example: what is the relationship between the soul-spiritual and the physical-bodily aspects of the human being? They have developed all possible kinds of theories about this. Here we have three types of theories. The one tries to come from the soul-spiritual and try to define this in some way, to formulate an abstract concept and then to try in how far the soul-spiritual can work on the bodily-physical. Another, more materialistically coloured theory assumes that the bodily-physical should form the basis, so that the bodily-physical brings forth the soul-spiritual as one of its functions. A third theory is that of psycho-physical parallelism, which assumes that the soul-spiritual and bodily-physical matter equally, and only to pursue how the functions of the one takes place beside the other, without looking for any exchange taking place between them. These are all psychological speculations. At the moment a practical situation is present, through psychology, through soul knowledge, it finds driving forces into the pedagogic didactic impulses.

One can simply say our sphere of observation of the soul-spiritual human being has not yet reached a principle which we are accustomed to follow naturally according to science. In natural science for example, when we look at the phenomena where warmth appears, without the usual kind of warmth coming forth as usual, then this warmth from other circumstances is considered as so-called latent warmth and how it had developed out of latent conditions, and now appears as warmth. Such principles which are common practice in science must — obviously metamorphosed in corresponding manner — also be taken up in the observation of the totality of the human being, in which the soul-spiritual is included.

One comes to such an approach of observation which is fully justified by science — if it hasn't yet been seen today — if one focuses on the first important transformation which happen in the human organisation with the change of teeth around th seventh year of life. Such transformations in the human being are usually only observed outwardly. However, the change of teeth is something which penetrates the entire human life deeply. Whoever trains his abilities of observation will learn to recognise how with the change of teeth an entire change in the child's soul life takes place. He learns to recognise how the child in the fullest sense of the word, didn't really live "in himself", but completely absorbed his soul-life in his environment. He learns to perceive the most essential of the driving forces in the child organism before the change of teeth, which is imitation. Through imitation the child learns movement.

One can through unbiased observation determine precisely how the movements from the father and mother or others in the child's surroundings enter into the childlike organism itself. One can follow how under healthy conditions speech is learnt under the influence of imitation. One can see how the child, in the fullest sense of the word, comes from his surroundings, with his whole being. This alters completely with the development at the change of teeth. Here we see how forces develop in the child, enabling the child to bring forth independent imagination, in which the inner child up to a certain degree is set free from the surroundings, which is not so before his seventh year of life. With the change of teeth, the child acquires a certain introspection and becomes gradually more accessible to abstraction.

Now the childish nature is again conditioned by everything which lives inwardly in the people surrounding the child, as it is absorbed by the child. That is why in the second period of life which begins with the change of teeth up to adolescence, is seen in such a way that everything which develops in the child is an adaptation of the people surrounding it. Not what the people *do* in the child's surroundings, because that is imitated, but in what *lives* in these people, this means what comes to expression in their words, their attitude, the direction of their thoughts, these are passed on to the child — as it were not through imitation but through taking up a power which is as part of him or her, as growth and nutritional powers in them: the power of authority. What is meant by the power of authority is not to be misunderstood because for someone who has written "The Philosophy of Freedom" it is necessary to point out how this authority principle comes under scrutiny in a certain phase of life. This means not the entire education should be put down to what is referred to today as the principle of authority. If one applies the corresponding value to such observations then things become ever more clearly differentiated and one gains the ability gradually to not only being able to observe the transformation in people from year to year, but also from month to month. What then happens actually between the change of teeth and puberty?

When you direct your gaze in order to learn what really happens there between the seventh and fourteenth year — those are of course only approximate numbers — the hidden forces within the child's soul now come to be expressed outwardly. This is hidden in the bodily nature and activates the expression of the human organism, works also in the formation of the brain in the first years of life and in the preparation of the speech organs, works also in everything the child develops in his bodily nature. Thus, you can say the following. Just as for instance the warmth in a body is hidden and can become free under certain circumstances, so the soul-spiritual, which works latently in the first seven years in the physical organism, expressed in every single movement, in every bodily process, only becomes free later. After the seventh year of life the body is left to itself more; the soul-spiritual does not withdraw completely out of the bodily, yet it does to a high degree. The change of teeth is then a kind of termination point of the first developmental phase, where the soul-spiritual was still in harmony with the bodily-physical.

You see that through this manner of observation you can reach a position where you are able to recognise a real relationship between the soul-spiritual and the bodily-physical. People don't theorize only around the question of how the one works upon the other, and so on. People simply see the soul-spiritual during one period of life as completely in the physical — this is clearly seen in the child's development — and later, after acquiring freedom, the form appears. So a comparison isn't made of what had been understood as abstract concepts earlier, but the reality is followed in the process of the soul-spiritual in the bodily form during the various periods of life. This means, however, that that which in natural science had been openly researched through the outer senses is lifted up into the spiritual sphere.

If you were to enter more into the details, which Anthroposophy wants to do, to penetrate it and not remain stuck in superficial definitions, you would soon see what kind of a faithful continuator of the justifiable scientific thinking of the spiritual scientific anthroposophical viewpoint actually is. Then, however, when in this way you gain the world of concepts and ideas of human knowledge, then the accusations regarding the alienation of the world of ideas is solved by itself.

Dear friends! Anthroposophy is the last to be in opposition to big and important events particularly during the 19th century in the pedagogical sphere, presented by great educators of humanity on pedagogical principles. Anthroposophy fully acknowledges the existence of great, meaningful educational principles and does not stand back before anyone in the recognition of the great educators. Only, you have to admit nevertheless, with all great educational principles there is often a certain dissatisfaction today regarding educational practice, educational methods; the most diverse kinds of educational practices bear witness to the fact that this is so. Why is this so?

It is often just a result of the intellectualism of our time. This intellectualism results — more than one normally believes — in a particular hostility towards life, especially in the social areas. It breeds in relation to ideation actually only abstraction. The abstract has no life-forces, it is in a certain way the corpse of the spiritual and is experienced as such. Even in having the most beautiful principles in which you can almost glow with enthusiasm — as long as these principles remain abstract, they can't obtain any kind of favourable influence. Only when these principles are permeated through with spirituality, living spirituality, which merges with the beings of people, could these principles become practical. Thus, Anthroposophy doesn't want to propose new educational principles in an abstract manner; it only wants to be an introduction for pedagogical and didactic skills, for the implementation of the art of education and art of instruction, and wants to present what the most beautiful educational principles can't give: spiritual foundations for the practical implementation, for the inner talents of teachers, to work in the school and in education.

For this reason, the Waldorf School is not so geared — as is often believed — to take our world view as it is conveyed to grownups, and to stuff it into our children. As a result, we have to particularly stress that as far as religious instruction go, the Catholic children are left to the Catholic priests, and the evangelist children to the evangelist priest. We have only arranged free religious instruction for those dissident children and if these lessons had not been organized they would have no religious education. As a result, some experience of the religious feeling can be accomplished because those parents who withdraw their children from religious instruction, send their children now into religious instruction in which we make the effort not to lecture Anthroposophy but to present it as is required at that particular age of the child. So it's not about depositing Anthroposophy into the childish mind, but it comes down to the teachers working through Anthroposophy, the pedagogical didactic methods employed in such a way that they really fulfil true human education.

This results in, simply through the practical implementation of such education and such teaching, not only in the child being looked at but the whole person being considered. It would be highly foolish to take the feet or hands as they are at a child's age and regard them as something complete and force them to remain as they are in childhood. It is obvious that we consider a child's organism as something coming into being, which has to be different later in life. However, in relation to the soul-spiritual we don't always do the same thing. We often even see rigid concepts introduced and that the child is frequently taught from a young age as having something like sharp contours in its soul. This is false! With anything which we want to allow incorporation into the childish organism, it must be introduced in a growing way, that it can gradually be transformed so that the human being later, in his thirtieth year for example, not only has a memory of what he had absorbed in his childhood but that the content of this has been as transformed by him as he had transformed his limbs. Everything of a soul-spiritual nature we give the child must also contain powers of growth, powers of transformation; that means we must make our teaching more and more alive.

Certainly this could be expressed as an abstract principle but practically it can only be accomplished when true intimate human knowledge is present. Such a kind of intimate human knowledge makes it possible to deduce everything from the childish nature into what is understood as the syllabus and goal of learning. Out of this the Waldorf School has taken its syllabus and the objectives of learning from actual human knowledge which can be read from month to month in the developing childish nature itself. The effort has been made to bring all of this about in a living sense.

I only want to mention one thing. Today in various ways teaching has improved even in some public tuition. But, you all know that during the school year the child becomes even more conscious than one is aware, and suffers under a system where the progress of the child is judged. It depends on the one side on the child's performance and on the other, the teacher's judgement of this performance, and is expressed as: "satisfactory", "nearly satisfactory", "nearly not satisfactory", "less satisfactory" and so on. I have to confess to you, I was never really capable of differentiating between "nearly satisfactory" and "nearly not satisfactory" and so on. With us in the Waldorf School it involves that out of the totality of progress made by the end of each school year, the child is given a kind of witnessing presented by the teacher which characterises each individual child and that he simply writes this on a piece of paper as his experience of that child. So the child sees a kind of mirror image of himself, and this practice — which doesn't depend on "satisfactory", "nearly satisfactory" and so on for the individual items — has been accepted with a certain inner satisfaction and received with joy, even when there is blame. The child also receives a kind of powerful verse which echoes with his own nature, which he takes up and which serves as a mission statement for the following year.

In this way one can, if one has the love for it, enter in a lively way into education, even working through unfavourable relationships in a lively way.

As a result, we come again to something which needs to be overcome, needs to be conquered in pedagogy and didactics in our epoch. Today one will hardly find any evidence in the outer historic descriptions, of how souls' constitutions have changed during the single evolutionary epochs. Whoever is without bias can readily understand how the spiritual utterances which were revealed to souls during the 10th, 11th and 12th Century for instance, are of a completely different character than what had been presented since the middle of the 15th Century to the soul constitutions of civilised humanity. Yes, up to the 20th century intellectualism in humanity has developed up to a culmination point. Intellectualism has the peculiarity, that it — just like the principle of imitation or authority — only shifts at a particular human age out of its latent position and in the case of intellectualism it is related to a later period in life. We see how the human being only when he has reached puberty, actually even later, becomes suitable to progress towards intellectuality. Before this age intellectualism works in a paralyzing, deadening way into soul activity.

As a result, we can say we live in an epoch which is only appropriate for grownups, which has as its most important cultural impulse, something which should only come into expression in adults. As a result, because our entire cultural tone is set towards grown adults, we are actually unable to understand the child — and even young people!

This is the most important aspect our civilization needs to look at. We need to be clear that precisely through those powers which our sciences and our technology have triumphed and have been brought to such a great blossoming, we must take up the possibility to fully understand the child and enter into the human nature of the child. It just needs our own effort to strike our bridge across to the young people and the child. What appears in various forms as the youth movement — one can say whatever one likes — has its deepest entitlement; it is nothing other than the cry of the youth: 'You grownups have a civilization which we simply don't understand, when we bring our basic natures to it!' — This bridge between the adult and the child's world must be discovered again, and to this Anthroposophy will contribute.

When you go down from the general cultural point of view to the individual you will once again find that these syllabi which are deduced from the essence of the child itself, teach us what syllabus we need to develop for the phases in the child's life. Reading and writing were in earlier times something quite different to what they are today. Take for example the letters: they are something abstract, strangers in relation to life actually. If we go even further back, we find something in the pictorial writing which is directly related to life. We often today don't even think about how intimately life was connected to this image rich writing and how strange these are in life: reading and writing. Yes, we stand within a civilization in which it is natural to have the strangest elements in life developed into civilization's goals. When we in an open-minded way look at for instance a stenographer or a typist sitting at a typewriter, we know that with such activities humanity has been sucked into the strangest civilization.

My dear friends, we don't want to be hostile to culture or become reactionaries, when we express this. Nothing is to be said about these means which have entered in modern times; they must be there. Yet, powers of thought need to be developed which can heal this once again, this, which if it is left to work all by itself could only lead to a definite decline of culture and lead to decadence. The most important moment in which a healing remedy can be introduced, lies in education and in the classroom, to be designed through education.

When the child enters elementary school, then it is indeed so, that the intellect is drowsy. The ability for abstract thinking, which first needs to be experienced through others, only appears later. As a result, abstract forms of writing and reading introduced to the child as it arrives in school, cannot be related to. We can only take something which can reach the child in a lively way, which works in the child itself already as an artistic soul principle, something more pure and splendid than any other art. This works on a subconscious level. We must continue this way and try to find forms of a particular nature, through which the child in an artistic way can be active in his total being in the artistic form of writing which can evolve into reading. With relevance to pedagogy, when the children haven't learnt to read or write at the age of nine or ten, one must have the courage to say: 'Thank goodness that these children can't read or write yet!' — because it isn't important for the child to learn this or that but that he or she learns in the right way at the right age.

This is why the Waldorf School education is orientated in an artistic way. Out of pedagogic-artistic principles it commences and gradually leads over into the intellectual. We take into account that music must appear early in education while this has a relationship to development of the will forces. As a result, we take into account that the usual physical education, as animated gymnastics are given as Eurythmy, is inserted into the lessons. It needs to be metamorphosed, transformed pedagogically-didactically, then those who have observed it discover that through this art of movement, the soul and spirit have been provided with something meaningful. One discovers that the child in his school-going education experiences him or herself into the art of movement in a similar way as a small child finds its way into speaking, with inner pleasure and inner naturalness.

Working from an artistic basis results in the child handling colours from very early on. Even though it is also sometimes inconvenient and might mean more stringent cleaning rules need to be applied, it will still affirm that the child enters more deeply into life than otherwise. Brought into the bargain is the development of a sense for life, that life doesn't go by but that the child lives with the outer world, that it becomes sensitive for everything which is beautiful, every encounter in nature and in life being meaningful. This is more important than the transference of details from this or that sphere on to the child.

Added to all that I've only indicated in an outline, the Anthroposophical foundation is what flows into the teacher's mindset, what the teacher simply through his entire being attributes to the pedagogical-didactic imponderables, when he closes the classroom door behind him, when he steps in front of the children. Whoever looks in a lively sense — not with abstract ideas — how the child copies and adapts to his environment, knows what works in the child in a soul-spiritual way. The teacher gets to know the child and as a result obtains the requirements with which to judge in quite a different way than is usually done. I want to present an example of this.

You learn quite a bit when you look at life in the following way. Once parents came to me and said their young son, who up to that point had been quite neat and tidy, had suddenly stolen something. I asked: "How old is the child?" The parents answered: "Five years." I said: "Then you must first examine what the child has actually done because perhaps he has not stolen anything." — What had he done in fact? He took money out of the drawer in the same way his mother takes money every morning when she wants to go shopping. From the money the little boy had bought treats which he didn't keep for himself but had given to other children. In this case a person can say: There is no reason to see this as stealing; the child simply saw what the mother did each morning and felt capable of doing the same. The child is an imitator. Each relationship of a child to the norms of adults, in which the expressions "good" and "bad" appear, only become applicable when the change of teeth has taken place. Therefore, we must obtain a completely different way to form judgements and learn that everything we do in the child's surroundings need to be so orientated that the child can copy them, can imitate them right into the imponderable thoughts within them. This proves the reality of thoughts. Not only our actions but also the manner and way of our thoughts give substantiality. In the child's surroundings we should not give in to any random thought because this works in on the child. Therefore, we need to look into even the imponderables in thoughts.

If one looks at how the child up to his seventh year lives in his environment one can get the impression of what the child had been before he came down into the physical sense world. Up to then — this is shown in anthroposophic research

— the person is surrounded by a soul-spiritual world which is permeated by the cosmos, just like in the physical world his body is connected to the physical world. We become able to see that in the child's life, up to his seventh year, it has been a true continuation of life before birth or conception. This however must be transformed in the pedagogic-didactic experience so the teacher, standing in front of the children, must say to himself: The super-sensible worlds have given me something to unravel, which I must level out in the path of my life.

Teaching and education really becomes an act of sacrifice towards the whole world. There is a conviction being uttered about teaching and education being a force and without which in real teaching and real education, nothing can come about. This conviction which hasn't been adopted from outside, but has come through inner work, through the anthroposophical world view, this is the most important in pedagogical-didactic work. You stand with shy religious reverence to what is hidden within the child's body, you look at one who has risen from eternal world foundations which is gradually revealed in childish movements, gestures and so on, and you know that the riddle of life needs to be solved in a practical way. Only in this way are the entire teach, and educational convictions directed correctly. This atmosphere which spreads in all activities, which needs to take place in the school life, is what Anthroposophy above all wants to have within the teaching and educational being and from where all details need their direction. However, to be master of them, it is necessary that you, through true inner observation of the smallest movement of the child's life, see how the spirit works right into its very fingertips. The teacher will acquire an inner overall view so that he out of an ability, which must become an instinct, meets his class in the spirit and skilfulness that come from his internal processing of the anthroposophical world view.

Here are a few indications which I was able to give; they could be implemented further in the next lectures. These indications should show that Anthroposophy doesn't want to be radically against great pedagogic accomplishments but that it will be the assistant to the great one, if we are not to remain stuck in abstractions, so that we can enter practical life in a vital way, in order for the art of education to become a real impulse, an effective factor in our social life!

"ANTHROPOSOPHY AND SOCIAL SCIENCE."

Lecture Five by Rudolf Steiner given in Berlin, 9 March 1922

M y dear venerated guests! Besides the introductory words I want to say regarding today's task, I want to limit myself to essential indications in the following lectures to details of the economic life in its relationship to the area coming under discussion.

Social science can't be talked about today from only a theoretical standpoint. Today — I mean at this very present moment — one can only speak about such questions while the dire situation of the economic life existing in the civilised world is in the background. Into this desolate situation was also added something which I set out in my "Key Notes of the

Social Question", after the temporary end of the terrible catastrophe of world war.

At that time, I urged everyone to observe the social economic life in relation to the present time of world development. It is this economic life at present which is intimately intertwined with that which moves in the entire circumference of the social question. Yes, most people at present can hardly sense that the social question can be separated from the economic question. Yet my book "Key Notes of the Social Question" [See Die Kernpunkte der sozialen Frage] enters into establishing clarity in relation to the area in question here, where it will be pointed out how the economic life within the social organism needs to establish its own independent position, such an independent position within which the same facts and indications acquire their form only according to economic principles, economic opinions and impulses. In this respect my book — I say it here in quite frankly, because that is what matters most — contains an inner contradiction. Only, this book is not to be regarded as a theoretical book on social science. This book wants to give suggestions above all to life practitioners; this book was written out of observations of the European economic life over decades. Because this book strives to be completely realistic, a direct encouragement for practical activities — practical action in the moment — it had to contain a contradiction. This contradiction is namely nothing other than what permeates our entire social life and consists in our social life being in the course of modern time mixed up, chaotic; only viable if it develops its individual branches from out of its own conditions.

I must speak about the threefold divisions of the social organism which leads to the economic life becoming separated in a fully, free way from the organised legal and state life as well as from the spiritual life, so that the economic life becomes, for those who stand within it, formed out of their personal actions and initiatives. However, we presently live at a time in which such a situation doesn't exist, in which the economic life stands within the structure of the general social organism. We live in a time in which contradiction is a reality. As a result, a manuscript, which has aimed at being written out of reality and is being offered with suggestions based on reality, can bring about a contradictory turn; it could only come from the standpoint of bringing the contradiction to clarity, and with this clarity achieve relationships.

I am thus in an unusual position today by giving this introduction because in connection to what is based on anthroposophical grounds, created with anthroposophical methods of thought, founded on decades of realistic observation of European economic relationships — it is in the widest circles where it was first misunderstood in the worst possible way. I can only say I fully understand these misconceptions which have been given to these underlying intentions; these misunderstandings are phenomena of our time. However, I must be on the other side of the standpoint, that in overcoming these misconceptions lies what we first have to strive for sociologically, socially, and to this I would like to say a few words to orientate us.

When my book "Key Notes of the Social Question" was first published, it took place in the middle of the European development which was immediately followed by the terrible war catastrophe. It was during the time preceded by the Versailles treaty, a time in which value relationships in central and eastern European states were essentially different. Not from some cuckoo land cloud impulse was my "Key notes" written down, but thought through from the immediate world situation in such a way that I hoped to believe a large number of people would find it, and on the basis of these suggestions search further, then one could — namely from central Europe — throw an impulse also into the economic development which would lead to a significant, acceptable ascent which from then on and up to today had been a continual waste on the economy and social life in general. At that time, you could say to yourself that a person could think out of this complicated world situation: Perhaps no stone will remain standing as he has created into the thought structures of the "Key Notes of the Social Question" —; that these ideas would be made up out of those who were there. Still, it could be grasped and would perhaps have given quite a different result to what could be fixed in a manuscript. It is not important that ideas are presented in a utopian manner, that an image can be presented as a social futuristic organism, but it comes down to people discovering and understanding: real problems exist here, directly in life; we have to deal with these problems out of our expertise and see if we can handle these issues by finding an ever wider understanding for them.

Basically, something quite different has happened. On the one hand theorists have all kinds of discussions regarding the content of my book, discussions to which all manner of demands are linked regarding its contents. Some theoreticians misinterpreted what had been said completely, wanting to turn it in a utopian sense by asking: How will this be, how will that be? — i.e.: what one could actually expect. It turned out to be a strange fact to me which took me by surprise because precisely those practicing economists who work routinely within the economic life, who know about this or that branch of business and rejected what I had said, spoke about things in their business which wasn't practiced in their business — that these practitioners argue over the key points of the social question and as a result, prove themselves to be abstract theoreticians. This shows that one can have a routine practical involvement in economic life — in the old sense; under the newer relationships it can no longer be — these practitioners were absolutely not in the position to what was being battered here as also being related to problems of the economic life, other than discussion points made in abstract theories; which could raise doubts when you oppose practitioners and get involved in their discussions because nothing concrete is entered into but only completely trivial generalities are repeated about the social question, if you question someone.

The other thing you could come across would be that at first those, who on the whole are quite substantial practitioners, even reject wanting to talk in this way about the possible form which economic problems could take on. Going on from here, some interest could be stimulated for instance in socialistic circles; here the experience could be that what is wished for is the least understood from that side and that everything should be judged according to whether they fit into old party templates or not. And so time passed by from when these suggestions were thought about. The whole terrible Valuta-misery came about which has to be considered in quite a different way to how it is usually judged today.

With the first appearance of my "Call to the German Nation and the Cultural World" and then "Key Notes of the Social Question", individual personalities immediately appeared who in their way are quite honest about healing central European economic life, and said: 'Yes, such proposals' — they called them proposals — 'look quite attractive, but it should first be asked how we can enhance the Valuta.' That was said during a time when the Valuta-misery according to today's relationships, still existed in pure paradise. Now it shows in such demands that tampering with only external symptoms are wanted. It has little understanding that Valuta relations battered on the surface show unhealthy economic relationships, that with such a cure of a symptom the evil is not addressed, and that it requires entering into much, much deeper social economic conditions today if one wishes to in some or other way arrive at speaking about problems realistically, regarding the indications in my "Key Notes of the Social Question." Now it has come about that what I repeated in conclusion of lectures which I held in the end of the "Key notes" at that time, had the call: people have to wake up before it is too late — that this "too late!" has come to the fore to a large degree today. We are not at all in the position to resonate in the original sense with the "Key notes" to understand them because in the meantime chaos has broken into the economic life where now quite other additions would be necessary and not what was merely mentioned but what had to be spoken about according to my conviction. One can hardly pass by a characteristic common to our age if one wants to discuss the most damaging aspect in today's economic life.

When I picked up the newspaper yesterday, I came across — and it could today be one of the most important symptoms we find everywhere, which our contemporaries express in single sentences — I came across the article "Postponing the resignation of Lloyd George until after the Genoa Conference". With this once again our daily situation is announced because the characteristic of today is "wait". "We want to wait" — this has actually become the ruling principle: wait until something happens but you can't tell what it will actually be. This is what is deeply embedded in the human soul today, on all levels.

Now I want to apparently — only apparently — introduce something quite abstract: this is intended in a complete realistic way because it indicates the forces working among us which have in the course of human development gradually enabled us to arrive at such a promising principle as "We want to wait" and apply it to everything.

When we look back at ancient cultural development we find in these old cultures, that factual thinking, in the sense as it appeared in ancient times — you know this from my lectures I held in the Philharmonic — can't be called purely "scientific". If one considers what stands in the place of today's scientific thinking then you will know that first of all, out of this thinking the economic life could not have directly emerged. The economic life had to more or less first become independent of human thinking, developed instinctively — not meaning automatically — as exchange in humanity. What wanted to be done in the economic life simply developed out of practical life. People acted instinctively; even expanded trade into this or that area but everything happened more or less instinctively. Now, one can from some point of view object according to today's understanding of human freedom, human worth and so on, to the economic conditions of olden times; all this would be good to be seen from the other side, how the extraordinary symptoms of human evolution, which even today can be instructive, for instance appear in the way employees and employers — if you want to apply a modern expression to olden times — lived in relation to one another during ancient Greece, old Egypt and right over to Asia. Today these things are taken in such a way that they elicit the sharpest criticism; but, each such a criticism is not historical, and one must say: the conditions in the corresponding epochs resulted from the feelings of humanity at that time. This is what one needs to focus on.

The other one is a fact connected to that shift in humanity's evolution which I've often pointed out, of around the fifteenth century, through which the soul constitution of civilised humanity became something quite different. I've already said outer history hardly points out that the collective soul constitution of humanity has become something different. If we ask ourselves how this human evolution relates to the economic life, then we get the following answer. The time for instinctive leadership as I've characterised, this time reached into the epoch of the shift. With this shift intellectualism arrived into the soul constitution; the drive to understand the world purely through human mental logic. This drive, which simply became a deep need in the human constitution, proved itself so brilliantly in the field of natural science and in that field, which developed as a result: the field of technology where in the most extraordinary way it has not celebrated enough triumphs. However, this intellectualism — it was shown in various arguments, which during this course have already been dealt with — has shown itself as completely incapable of understanding the phenomena of human life and human nature as well as social relationships. With this intellectualism, this intellectual orientation, the soul can be brought back in a grandiose way to outer sense perceptible nature and its laws. You can't intertwine the one with the other in this intellectualism and while this intertwining goes on, get organised and while organising yourself also enjoy life and grasp spiritually permeated social relations. I would like to say the following. The network of intellectual ideas is too broad for what lies in social life. To think scientifically — that, humanity learnt from this intellectualism. Everything has been drawn into it, even theology. Intellectualism rules while we observe and experiment with our entire scientific way of thinking, and finally, what we have introduced into it which can't fit between the lines of intellectualism, we see as not scientific.

During this time intellectualism fell into the transition from a purely instinctive economic life to one fuelled with human thoughts. We may say that in the time when people didn't think intellectually about the world, the economic life was directed instinctively. When however, the time came when more and more world economy and world traffic appeared, this tendency required human beings to penetrate world economy and world traffic with their thoughts. These thoughts only came from intellectualism. As a result, everything which came from economic thoughts — in mercantilism, physiocratism, in the national economic ideas of Adam Smith, as in everything which later appeared right up to <u>Karl Marx</u> — on the one side demanded economic life, which was not merely instinctively mismanaged but it was grasped with thoughts, however on the other side, where thoughts could only come from intellectualism, all economic observations would become thoroughly one-sided, so that nothing could result from this economic observation which could be seen as continuing to work in economic practice. On the one hand you have the economic theorists who created axioms from intellectual sentences — like for instance Ricardo, Adam Smith or <u>John Stuart Mill</u> — and who now develop systems on the basis of these principles on which they built a complete self-contained mentality (Geistesart). On the other hand, the economic practice needed and demanded penetration by the spirit but found no connection to what had continued to work instinctively and as a result it fell into complete chaos.

So these two streams became more and more common in recent times, on the one hand were the economic theorists — without the influence of economic practice; on the other hand the practitioners with their old practices which had become a continued routine and as a result let the economic life of the civilised world fall into chaos. Obviously one must express such things in a somewhat radical manner because then one will really distinguish what works and what can be understood as a problem.

If one now wants to find, I might call it, a connection, a kind of synthesis between economic thinking — which has gradually been eradicated by practice — and this economic practice — then one finds this connection at least in one of them. Recently a kind of economic realism has developed; a kind of economic-scientific realism which says that one can hardly find general laws for economic life if economic facts are not considered and events between single nations or groups are looked at what has happened only in an external way, to find guidelines for economic trade. From this basis has developed the so-called social-political in economic law-making. This means people gradually believed they have discovered through mere observation of factual economic relationships in connection with the permeating social connection that they could find certain guidelines which could be brought to expression in economic law-making; people now had, by taking detours through the State, tried to actualize some of these which had developed out of observation, but as a result it had to be actually admitted that these foregoing observations of real scientific economic laws could not at all emerge. Yes, we are actually still basically in this situation today. Just when one is in the situation of encountering decisive experiences, I could call it social Ur-phenomena being judged in the right way, then one sees the situation one is in.

You all know that Woodrow Wilson's "fourteen points" arrived at the dreadful chaotic point civilised life had entered.

What were these fourteen points actually? They were basically nothing other than abstract principles of an unworldly man, a person who knew little about reality, who appeared in Versailles where he could actually have played an important role. This man who was a stranger to reality wanted to show the world how to get organised according to principles founded on intellectualism. One only had to experience with what inspiration civilised mankind hung on to these fourteen points, however with the exclusion of a large part of the central European population, they unfortunately also fell for these fourteen points after a short period of time.

During 1917, by contrast, I tried to show individual central European personalities who were interested but who were not following it, but were either approached or brought to it, how abstract, how unrealistic this was which wanted to be brought into the social form, how so to speak everything which ruled in the poor educational principles in modern civilisation was a concentration of what this school master Woodrow Wilson had introduced, and how the abstract principles — in a bad sense — were received with enthusiasm. At that time, I tried to show that a healing of the relationships could be entered into if you take a stand in opposition to such abstract attitudes, without excluding thoughts but which promotes realistic thoughts in order to develop from a realistic basis. Then it will not be a utopian invention — I would like to say the Woodrow Wilson principles were the most condensed utopian, utopianism already in its third potency, but one must be clear about finding contemporary humanity in its real conditions in order to discover impulses. Therefore, I gave up having to deal with any utopian theory, refrain from even saying how capital, how labour and suchlike must be formed; I gave at most some examples for how one could think about forming the future according to contemporary relationships. That was however only as illustration to what should happen; because just as I have spoken about the transformation of capital forces in my "Key notes", just so this transformation can be fulfilled in a modified way. It is not important for me to present an image of the future but to say from which foundations, in a real way, one can now — not with theoretically thoughts — come to an actual solution for the so-called social question. It is not important to say that this or that is the solution to the social question. I have already had too many experiences in trying to find such a solution. Already in the 1880's in pleasant Vienna all kinds of clever people gathered nearly every afternoon after two o'clock. In the course of one hour the social question was solved many times! Those who considered the relationships of the present in an unbiased manner, know very well that solutions which often appear in thick books have much less worth than those negotiated in comfortable Vienna with a stroke or two of the pencil and fantastic words across a white tablecloth. That is not the point and it was the worst mistake brought to me that it should be something like that.

What I wanted to point out was the following. The solution of the social problems can only be affected in a real way out of itself; the result can't be solved through discussion but through events and actions. Conditions first need to be established to contain this activity, conditions I try to refer to in my "Key notes" and in other sources. I'm trying to show we need arrangements in our social organism which makes it possible for a spiritual life to develop out of its own conditions in which the spiritual life itself works; that we need a second member, where only legal-state impulses are at work, and besides that a third member, where only those impulses work which come out of production and consumption of goods, and lastly that it develops out of an associative economic system, culminating in healthy pricing. In this way the old class system will not be recalled. It won't be people branching into an educational class, a defence class and a nutritional class, but the modern human being has moved into individuation and will not be divided into some particular state.

What exists externally as an arrangement simply comes from the forces in history's unfolding, which are separate from the conditions out of which they are negotiated, to do something for the spiritual life, the legal or state life and economic life. Only when conditions are created which for instance the economist can do purely out of economic impulses, which would be modified by contemporary market trends, or should modify the capitalistic relationships, only when such possibilities are created among people will they develop something of a real solution — which is in a continuous becoming — of what can be called the social question.

It is not important for me that the social question is solved because I have to agree that the solution can't at any one moment be given as something self-contained, while the social problem from which it has originated is in a constant forward flow. The social organism is something which becomes young, and older, into which new impulses must flow, of which the following can never be said: it has this or that form. If the social organism is not so, that people sit together in one parliament that mixes all interests together, where those interested in economics make decisions about questions of the spiritual life, legal life and economic questions and so on, but when in a healthy social organism each individual sphere is considered out of its own conditions, then the state life can be placed on a realistic democratic foundation; then what is to be said doesn't come from one person in one such a single parliament, but it will emerge from continued ongoing negotiations among individual branches of the social organism.

In this context my book was also a warning to finally stop the fruitless arguments about the social question and to place it on the foundation where the solution to the social questions can be taken up every day. It was a call to the understanding how to take what was abstract in thought and to really translate this into thoughtful action. Added to this for example the associations can serve the economic life. Such associations are different from those which in recent times have been established as socialization and can be created every day out of economic foundations. They are concerned with those people who handle goods production, in the circulation and consumption of goods — which every person is — to unite in associations through which healthy pricing can develop. It is a long way from knowledge of the subject and specialised knowledge which have to be achieved by people linked to associations, up to what doesn't come from legislation, also not from results of discussions but results from experience, which will give healthy pricing. Above all people have the desire, the broad outline of what they want at the time and which I am trying to present to you to discuss through these introductory words, because the world is so schooled in abstract thinking that one also takes this suggestion only from the point of view of abstract thinking, which I'm only using as an illustration, and discuss it for hours, while it should be about really understanding how each day the members of the social organism can be tackled in the way as indicated in my "Key notes."

Today it is not of importance to find theoretical solutions to the social question but to search for conditions under which people can live socially. They will live socially when the social organism works according to its three members, just like a natural organism under the influence of its relative threefoldness also work towards unity.

You see, it first has to be explained what is meant by such things. When these things are spoken about, words are still required; yet words need to be taken up according to their intellectualised meaning which we attach to them today. These are translated immediately into intellectual things which are quite clearly not immersed in intellectualism. Therefore, in my book I have spoken in such a way about capital and about the natural foundations of production simply as ideas being thought out. When we want to deal with things abstractly, we can create definitions for a long time, and that has in fact happened. Someone says with equal right: Capital is crystallized labour, work which is stored up — and someone else says with the same right: Capital is saved labour. You can do this with all economic concepts if you remain within intellectualism. But these are not all things which can be dealt with theoretically only; we need to understand them in a lively form. If practitioners do a lot for the benefit of their practice and routine, cultivated out of the abstractions in these things, they can achieve the following, which I want to explain through a comparison.

I look at Ernst Muller. He is small with completely childish features and childish qualities. Twenty years later I look at Ernst Muller and say that this is not Ernst Muller because he is small and has childish qualities and quite a different physiognomy. — Yes, if at that time I had formed a concept of Ernst Muller and now want to attribute him with what at that time I had met as his real being, then I'll be making a terrible mistake. As little as people want to believe this, yet it is the way people are thinking along economic routes. They form thoughts and ideas about capital and labour and so on, and they believe these ideas must always have the same validity. It is not necessary to wait twenty years; you only need to go from one employer to another, from one land to another to discover the concept which you had created in one place is no longer valid in another because a change has been brought about — like in Ernst Muller. People don't recognise what exists when one doesn't have mobile ideas moving within life.

This is what makes it possible that on an anthroposophical foundation today's needs also find their expression in economic institutions because Anthroposophy's nature involves flexible ideas, which can teach you how you can provide your ideas with forces of growth and inner mobility and that with such ideas — as little as today's practitioners want to believe it — they can dive into other kinds of reality, which are revealed in the social life between one person and another, between one nation to another, through to entirely what has become necessary now in the artificially impaired world economy. One can therefore rightly say it is not an external attempt made on Anthroposophical grounds towards social ideas but to arrive at social impulses.

I still remember a time when many discussions took place about these things. I always had to stress: I'm talking about social *impulses!* — This upset people terribly. Obviously I should have said: social ideas or social thoughts, because the people only had thoughts in their heads about such things. That I spoke about impulses angered them terribly because they hadn't noticed I used "impulses" on this basis of indicating realities and not abstract ideas. Obviously one had to express oneself in abstract ideas.

Today it must again be grasped that a new understanding must be found for what is called the social question. We live in different relationships today than in the year 1919. Time is moving fast, especially in economic areas. It is necessary that even those very ideas which were considered at that time as mobile, continue to be contained in the flow and that one's observations of viewpoints stay within the spiritual present.

Whoever wants to look at the reality of relationships within the economic life knows they have essentially changed since the writing of my "Key notes" and one can no longer just use deductions as before. At least in the "Key notes" one would find an attempt to search for this method of social thinking in a realistic way, perhaps exactly because this attempt has grown from the soil where realities are always looked for, where one doesn't want to fall into fanaticism or false mysticism — because this attempt is grown out of accuracy on the wrestling ground of the anthroposophical world view.

"ANTHROPOSOPHY AND THEOLOGY."

*Lecture Six by Rudolf
Steiner given in Berlin,
10 March 1922*

M y dear venerated guests! As an introduction I have been obliged to refer to a notice in the newspaper which has just been handed to me; a notice in "Christian World," a publication I don't know and obviously have not thought about. In this notice it says: "From 5 to 12 March an Anthroposophic University Course will take place in Berlin. The day for theologians is Friday the 10th. This event on Friday is now an unequivocal challenge of Steiner and his followers to the theologians ..." and so on.

Now, my dear friends, this event may be anything; what it certainly isn't, even if it was believed to be, it would be misunderstood in the most profound sense, if it is regarded as a challenge to the theologians. I myself would not be involved in any other way than having been asked to cooperate through lectures and introductory observations in this university course which didn't come out of my initiative. I'm least involved in today's event (which is an insertion into this program item of the course) by thinking that what we were dealing with today could be understood as an "unequivocal challenge of today's theologians."

Thus, you will also allow, my dear friends, that not all sorts of misunderstandings will again be linked to what I have to say in a few introductory words today. I want to limit myself to a theme: The relationship of Anthroposophy to Theology. I want no new misunderstandings to arise; I will renounce some of them in my presentation because otherwise I would have to once again find my intention misjudged.

Dear friends, it has never been my purpose — forgive me if I'm forced by this challenge given to me by shortly mentioning some personal details — it has never actually been my intention to challenge theology and from their starting point Anthroposophy had, insofar as it presents a work sphere in which I participate as well, never attempted to set them apart within the work, with today's theology. This has happened so far, and really from me it has happened as little as possible, but unfortunately it has resulted that many attacks against anthroposophy from the side of theology have taken place, and sometimes people — not me particularly but others — defends themselves. Anthroposophy wants to remain thoroughly neutral in its working sphere, I'd like to say, it wants to work out of present day spiritual science.

Towards the end of the previous century one had a certain scientific direction, certain scientific methods, an attitude and method, out of the foundation of which we have already spoken, and which can't be spoken about more extensively, established a method and attitude which people apply to the entire development of recent times and particularly apply to scientific research. Through this natural scientific research, the greatest possible triumphs — I don't mean in a trivial but in a deeper sense — have come to human progress and human well-being. During this time natural scientific research stands in a somewhat puzzled manner towards philosophy. Philosophy had to separate itself from those methods which are applied to natural science; the difference of a factual sphere made scientific methods inapplicable in philosophy.

People were not always, one could call it, theoretically and epistemologically clear in what sense the scientific methods or philosophic methods had to apply. Practice lapsed into experimental philosophy in certain areas where it was more or less apparent or more or less really worked, but the uncertainty is basically there as well. By contrast Anthroposophy worked out of the most varied foundations towards its own working methods. On the one hand it wants to take into account what can be achieved in modern thinking and research methods of science, and on the other hand the human needs for the spiritual world and its knowledge. The human being is confronted on the one hand with the fact of fully recognising scientific methods, and in relation to the treatment of the scientific field — I have already mentioned this — I am today as much a student of <u>Haeckel</u> as I was in the 1890's; not in the sense of scientific methodology not to be developed further and not as if, from the side of science Haeckel's writings should not be applied, but it comes down to quite a different area being discussed. In the treatment of the purely natural world I'm as much in agreement with Haeckel as at that time. It deals more with the experience of natural scientific observations through which one is educated in scientific precision, in a natural scientific sense which can result in the creation of ideas and concepts, which are needed for working scientifically. This then holds true for all observations in the world — due to our limited time now, I can't give you proof of this. This remains a truth: for all outer sensory observations this sentence is valid: "there is nothing in the mind which wasn't previously in the senses" — certainly on the other hand, Leibniz's statement applies: "Except in the mind itself."

In the experience of the mind, that means in the weaving of the soul through the mind's categories where ideas are experienced in objects of nature, the examination of facts of nature which need a formulation of natural laws, in which experience of the world of ideas live, there is something which goes beyond the mere sensory experiences, so that when a natural scientific researcher confronts natural science, he must say to himself, if he is sufficiently unprejudiced: everything in the mind must be created out of the senses, only the mind itself can't be created out of the senses.

Once you have understood this in a lively manner then there is no obstacle to now observe what inwardly to some extent can be looked at in the pursuit of the expansion of the mind's categories through an inner soul-spiritual process, through such a process which is inwardly quite similar to the outer growth processes seen in the plant and animal. One remains always true to one's conviction of natural development when one admits that out of the seedling, if you have an inner image of it, you gain a truth which is that the mind itself can't be created out of the sense world. One remains true to that which is learnt from natural existence when you make an attempt to observe the human mind as a seedling which can grow within. When you make this attempt in earnest then the rest is a direct result of what I've suggested here and in other places, of the growth of human intellect in Imagination, Inspiration and Intuition. This is simply a fact for further progress in inner human development. Through this the result is a true observation of the spiritual world. This observation of the spiritual world Anthroposophy tries to clothe, as well as possible, in words of today's language use. Naturally one is often forced that what one is observing — I admit this without further ado — is clothed inadequately in words from the simple basis that speech, as in all modern languages, in the course of the last centuries adapted to the outer material world outlook and today we have the experience, which we have with words, of already being more or less orientated to this world outlook.

As a result, we always struggle with words if we need to dress in words what we have observed through Imagination, Inspiration and Intuition in such a way that it can really be proven again through the ordinary, healthy human mind, because this must also be a goal for Anthroposophical research.

So Anthroposophy was simply a field of work and as such a field of work it has become, in the strictest sense of the word, conceived by me. Those individuals — and they make a very small circle — who have the need to hear about such research methods in the supersensible world, will be told and shown what can be discovered in this way. Nobody in this Movement will be forced in any way to participate in something other than through their own free will. What is said about this, that some or other suggestive means is applied, with one person it is a conscious and with another it is an unconscious defamation of what is really striven for in the Anthroposophic Movement. It is true that whoever thinks it over with a healthy mind, what is researched in Imagination, Inspiration and Intuition, in his higher senses becomes a more free person than any other people living in the present. His contemporaries for instance follow currents in parties and are influenced by all kinds of suggestions. From this inner soul dependency Anthroposophy must free people, because it claims that everyone, who wants to live into it, will not merely become immobilised in simple passive thinking, but that this thinking will make them inwardly mobile and powerful, and this empowered thinking makes a person more free.

For reasons, into which I don't want to enter today, it happened that from the scientifically orientated people on which Anthroposophy actually depend, in the beginning only very few drew closer to Anthroposophy. Today we have really made a start. Those people who first entered into the Anthroposophical Movement — with more or less naive minds with strong soul needs — they were never told anything other than what could be found in a conscientious way within anthroposophic research. I'm always delighted when things are said to me, for example by one of those present here today, a very honourable personality: 'It is actually remarkable that you even get a large audience, because you avoid actually talking in the way which is considered popular, which we call understandable. You speak in such a way that people actually always have to do work to listen and this people don't want these days, so one must actually wonder how you still manage to find such a large audience.' — These are what the words sound like, which I've heard for years and now a seated person here has also said them, after they had heard a course of my lectures at that time. For popularity I have never striven because I have the validity of Anthroposophy which I want to bring to the world.

Now it is extraordinary that people from all kinds of circles of life and circles of commitment have come. Because Anthroposophy came their way simply through their work in a certain relationship to religious streams of the present, it actually never came into conflict with religious needs of people who came to it: to people, like I said, from all walks of life. For instance, I have often been asked by Catholics who find themselves in our midst whether in connection with religious practice it would be possible to remain Catholics when they also take part in the Anthroposophical Movement.

With Catholics I must say: Obviously it is possible for a good Catholic to take part in what Anthroposophy has to offer because Anthroposophy is there, not to limit the knowledge which speaks about the supersensible world, but it forms a foundation on which supersensible research can be done. This is my preference, that what comes out of the supersensible world is spoken about without entering into any kind of polemic. Someone who honestly says what he sees, knows how polemic comes about and how unfruitful that really is. My original striving was simply to honestly say what is found through Anthroposophy and to exclude any polemic considerations. Things don't always happen this way in life. Still, within the Anthroposophical Movement people of all faiths are found together, and so I would like to say that Catholics may obviously take part in the Anthroposophic

Movement, but it will only come into one single point of conflict in the practical religious exercises and that is the audible confession. Not on the basis of it being an audible confession because that could be considered as a matter of conscience. I have found enough protestant clergymen who have gloated over a kind of confession in order to develop an intimate relationship with the congregation. One can have various opinions regarding this. However, here the point is that the Catholic Church denies the altar sacrament to anyone who has not made an audible confession before it. Due to this impediment, taking part practically in the most important Catholic church sacrament is difficult because those beliefs which are gained from the supersensible world need to be combined with this behaviour which is not freely done but which have nevertheless to be adhered to in the Roman Catholic Church constitution. The audible confession, as it is handled, tears the Catholic away from freely following the supersensible world, not because of Anthroposophy but because of the Roman Catholic Church constitution.

This could be avoided if confession could be avoided. One can't avoid it because otherwise one can't participate in the communion service. Still you can find many Catholics who search within the Anthroposophical Movement to satisfy their soul needs.

My dear friends, it is of course natural that people of all beliefs come to Anthroposophy, it is natural that simply in our time a strong need has developed to express what

Christianity is about within the Anthroposophical Society. Now I would like to say the following. Just as with all other phenomena of research, in as far as the phenomena of the supersensible and sensible world flow together, just so Anthroposophy regards the content of Christology; it likewise tries to help with research into the supersensible regarding the content of Christology, help which can be acquired through anthroposophical methods. Now it is difficult to say in only a few words what characterises the position of Anthroposophy regarding Christology, but I would like to say the following.

We observe people in earthly life between birth and death where they have their soul and spirit life in their physical being, that they are bound to their physical body in relation to what they observe and process whatever is presented to them in their environment, also in relation to work itself, in relation to their life of will and finally in the way in which they place themselves in the sensory physical world. When a person looks back at when he wakes up, naturally in his surroundings, he firstly finds perceptions possible through the senses of his body, through his mind, and all of these experiences and observations of his environment he experiences as combined.

However, because his mind, intellect and ancient spirituality are carried within his own spirit, so he can — if he only thinks enough about himself, if he only looks away from the environment and looks at himself — not deny that through his own activity he comes to the conclusion culminating in a concept which only has spiritual content and that this spiritual content — if I may express it this way — is the Father-godly imagination. Here anthroposophical research must be of help with its methods. I can only briefly characterise this. It makes the entire human cognitive work process clear — this will also emerge out of the lectures in this course. It also wants to point to what happens through people when they try to turn their gaze away from the outer world, in order to gradually observe their own past actions and ask themselves: What have you actually done? What justifies you at all to make an imagination of the outer world? — By researching this experience far enough a person — when I may use this expression again — comes to a Father-godly experience. Whoever examines this divine godly-Father experience through Anthroposophy, arrives at quite a definite judgement. I ask that this judgement, which is a fact, which I speak about radically, should not be misunderstood.

A person arrives at this verdict, a person who is totally healthy — totally in full health in his physical body — comes to this godly Father experience, this means that whoever doesn't arrive at this godly-Father experience carries some or another degenerative symptom, even if hidden. In other words, through Anthroposophical research you can say: To not come to a Father-godly experience indicates some human illness. That is of course radical to say because illness is ordinarily seen through physical means because — if I might say so — it dwells in the subtleties of the human organisation. In fact, it is clear to those who research through Anthroposophy: Atheism is illness.

What I've said yesterday about the development of opinions, right or wrong, this is particularly important here. If a person follows only this route, then he will come to a Father-godly experience. When he then goes further in this way, if he becomes aware what shortcomings live in his soul, if he only comes to this Father-god experience, he becomes aware that basically in the limitation of modern humanity leaning towards intellectualism there also lies a kind of limitation of this godly-Father experience, then he will realise he must go further with this godly-Father experience. Here outer observations can support this easily.

It is an extraordinary fact that in western countries where natural science has grown to its maximum intensity and where this scientific attitude doesn't want to enter into discussing the supersensible, but that religion must remain preserved, that just in these religious movements of western countries the spirit of the Old Testament has particularly and successfully intervened even in our modern time. We see how in the west, when Christianity is outwardly accepted and preached that it is done totally in the spirit of the Old Testament; in a certain sense Christianity reshapes the Father-god and doesn't discern a difference between the Father-god and Christ.

In the (European) east by contrast, where people's minds don't see the division between religion and science as sharply as in the west; in the east where this bridge for the human soul more or less exists as an elementary inner soul experience — we find that for example in the presentations of the great philosopher Vladimir Soloviev — how the Christ experience, as an independent experience, exists beside the Father experience.

In this way one can say to oneself: indeed, a completely healthy person can't be an atheist if he combines everything around him in the outer world into the culmination of a God-imagination, which he must give a spiritual content; yet he remains with only a Father-imagination. With this Father-imagination one doesn't arrive at a summary of outer natural phenomena, it fails immediately when applied to one's own human development; one is then, as it were, abandoned. By deepening this inner development from this point at which one has arrived, having taken up the outer world into one's soul — then by following this inner development one will, if by open-mindedly pursuing it, come to a Christ experience, which is initially present as an indefinite inner experience. This experience continues to be recognised by Anthroposophy. A person, simply through honest observation of the human evolution on earth, comes to seeing before his own eyes, the Mystery of Golgotha, the historic Mystery of Golgotha. He arrives here through the inner development of spiritual organs which direct him to

Imagination, Inspiration and Intuition. If one with the help of these research means pursues the way human development went from antiquity to the Mystery of Golgotha, then one finds that everywhere in religious imagination — not only in the Old Testament religious imagination — lived a gravitation to the coming of the Christ-Spirit.

Then one can simply through observation, learn to recognise how the Christ-Spirit was not united with the earth in the time before the Mystery of Golgotha. By pursuing all of this which was sought for in the mysteries, was popular in pre-Christian religions, then we see how the images they made of their gods, finally all melt together into what the Christ-Imagination is. We see how the minds of people all over the world are lifted to the supernatural when they turn to their gods in their souls. We see how the point of origin for earthly mankind's development was simply more given through the human organisation than what was perceived through the senses or the mind in what could be observed in his surroundings. It entered into the human soul — most strongly in ancient times, and then less and less — what I would call instinctive perception — not earthly — of the world, to which the human being felt he belonged. In the moment when a person, through the mysteries or through popular religion, is brought to where he can lift his soul into seeing extra-terrestrially, and with which he knows he is united in his deepest being, at this moment a person experiences a rebirth within himself.

Now my dear friends, when we follow human evolution from an Anthroposophic point of view up to the Mystery of Golgotha, it shows that these abilities, which dwelt within human beings, actually diminished gradually and were no longer there the moment the Mystery of Golgotha took place on the earth. Certainly there can be remnants, for evolution doesn't take place in leaps. Individuals preserved, though perhaps inaccurately but still instinctively, an awareness of what had once been seen; this can be pursued in art. Then the Mystery of Golgotha took place on earth. In the Mystery of Golgotha Anthroposophy sees the streaming in of that spirit which previously could only be searched for in the extra-terrestrial: the in streaming of the Christ into the human body of Jesus. How this can individually be imagined, can only be discussed with those who have engaged positively in these fields of research. Here Anthroposophy shows how from that time onwards, from the time of the Mystery of Golgotha, another time has begun on earth, a time about which all the old religious knowledge confessed about. The Christ who went through the Mystery of Golgotha, the Christ who Paul saw on the way to Damascus, the Christ then remained within in the earth with humanity. This is what these words want to say: "I am with you every day until the end of the world." He lives among us, He can be found again. The Paul experience can, with certain preparation, be renewed time and time again. Then, if Christ is searched for in this way, a person — by looking at his own inner development — just as since the

Mystery of Golgotha happened on earth — can see Christ walking; he discovers Christ in his inner life in the same way as when in the outer world — if he is not ill with atheism — he found the Father-god.

Thus, I can only fleetingly, in a sketch, indicate how Anthroposophy through real research of the Christ event, can arrive at an inner objective fact. With all possible detail Anthroposophy tries to present the Christ event as the most important fact of the earthly life of humanity, as something which happened objectively. For this reason, the entire spirit through which the Christ event is presented in Anthroposophy is done in such a way that this event can be absorbed simply as fact. We have within the anthroposophic movement experienced that for example Jewish confessors found themselves in the most genuine, truest and honest sense in recognising the Mystery of Golgotha. With this, my dear friends, the Anthroposophical Movement has already anticipated what after all must enter into human evolution: through directly pointing to what can be seen in the Mystery of Golgotha, how the way to Christianity can be found again.

There is always a question whether there isn't yet a deep meaning in the book by Overbeck, a friend of Friedrich Nietzsche, that modern theology is no longer Christian. If this is legitimate then one could even, perhaps with a certain right, say: Anthroposophy is suitable for directing people in a lively way to the Christ experience. It states that during the time in which the Christ event took place there still existed an instinctive insight among some individuals, so that the spiritual foundation, or I might call it, the spiritual substantiality of the Mystery of Golgotha could be seen and acknowledged in the first Christian centuries. We then see how this diminished gradually; we see it completely fade in the figure of Scotus Erigena, we see medieval theology spreading where the attempt was being made to separate itself from what modern humanity had to develop in the intellect, that which, when it is left to the person who no longer develops inwardly, he becomes incapable of accessing the supersensible worlds. It split what wanted to enter into the human soul into what was recognisable by the intellect, and what people could not attain themselves, except through a revelation.

On this basis one can understand the entire medieval theology, especially Thomistic theology which was considered by Catholicism as the only authority. Today something can be said about this. What Anthroposophy was and is, is nothing other than simply to express what exists and is available through spiritual observation.

As Anthroposophy comes to the proposition that atheism is actually a hidden illness, it arrives at a second proposition: Not finding the Christ, not finding a relationship with the Christ is destiny for humanity, is the fate of misfortune. Atheism is an illness, not finding the Christ is the fate of misfortune because one can find Him in an inward experience. Then He positions Himself there as that Being who has gone through the Mystery of Golgotha. One can only discover Christ through one's inner life; one doesn't need anthroposophical research to be a religious person in the Christian sense. Then again, when one has come to Christ, one becomes a member of the spiritual world and one can really speak about a resurrection of the human being in the spiritual world, because the person who fails to find Christ in regard to his world view, is restricted. Atheism is an illness! Not coming to Christ is a destiny, not reaching the spirit is soul obtuseness!

Now, my dear friends, Anthroposophy relates from such foundations basically only to religion (and not theology) and to religion only in as far as people who have religious needs and who are unable to fulfil them through current declarations, approach Anthroposophy. Anthroposophy will only do what is necessary within the needs of today, and that which others fail to do. What ethos is at this basis — I have to always characterise this again — you can find from the following.

Some years ago, I once held a lecture in a southern German town — at that time it was a German town but it no longer is — a lecture entitled "Bible and Wisdom". Two Catholic priests were present at the lecture. After the lecture they both approached me and said: "We actually haven't found anything in your lecture which could be challenged from a Catholic point of view."

I answered: "If only I could always be so lucky!"

To this they both replied: "Yes, but we noticed something, it is not *what* you say but it is the manner and way *how* you present it. We must add that you speak to people who are prepared in a certain way. You lecture to a kind of congregation who have a certain education; we, however, speak to all people."

I said: "Reverend, it doesn't come down to how our subjective experiences decide, but it comes down to us living into our work in evolution, that we don't imagine we speak for all people but that we answer such a question according to what objectively lives in the evolution of humanity. So, I can imagine I speak for all people — and could be very mistaken — you can imagine that. It is very good for enthusiasm to have such an imagination. Still, ask yourselves for once: do all people who have the need to hear something about Christ all come to church?"

Both of them couldn't say yes because naturally they knew that a lot of people who search for a way to Christ, do not come to the church.

So I said: "You see, for those who don't come to you and still search for a way to Christ, it is for those I speak."

This means finding your task in the evolution of time, and not to imagine you speak for everyone, but to ask: are there minds out there who want to accept this or that in a special way?

Anthroposophy never turns to any other mindset, like to some or other religious confession.

When we, in the Waldorf School, manage to apply teaching in a practical way out of Anthroposophy we still completely avoid making the Waldorf School a school which will splice Anthroposophy into the heads of the children. With regards to religious instruction, we leave the Catholic children to be instructed by a catholic priest and the evangelists by an evangelist priest. Only for the dissident children there is a freer kind of religious instruction, but in the thorough Christian sense. We don't introduce abstract Anthroposophy — also no concrete anthroposophy which is presented to grown-ups — but we try with all our good intensions to bring to the children what is suitable to the stage of their development; all of that must first be searched for and determined according to the content and method. Through those of us who have given free religious instruction, we have managed to bring those children who have no religious instruction as such, towards Christianity and they come in droves to take part in this kind of religious instruction. Never have we preached some or other kind of religious propaganda within the Anthroposophical Movement and even less would Anthroposophy embark on something against single theological systems. With this in mind, anthroposophy can only apply itself to finding differences in separate theological systems in order to understand them and not to oppose them. Thus, I've always regarded it to be my task when I speak to people who have come to Anthroposophy: to make it understandable why Catholicism has become Catholic, Protestants Protestant, Judaism Jewish and Buddhism Buddhistic and how all of them — I believe that is a Christian concept — have within them a Being who through their destiny will let them experience the true Christ. So it is not possible, if attacks have not originated from the other side, to start a struggle between Anthroposophy and theology, and also today I want to utter these words, while it has been asked for from those who organised today's theologian's day. The only task of Anthroposophy is the pronouncement of anthroposophic research results about the supersensible worlds. This is why I have always been reticent in particular regarding attacks originating from the theological side.

Anthroposophy doesn't want to act as a fighter on the scene but to satisfy the legitimate demands of human soul needs of the time. Everyone who in this sense wants to work together with Anthroposophy and wants to bring to the surface the fulfilment of legitimate, soul foundations of human soul needs, everyone who wants to work with her in this sense, is welcome!

"ANTHROPOSOPHY AND THE SCIENCE OF SPEECH."

*Seventh lecture of 7
given in Berlin, **at the
Singing Academy**,
11 March 1922 at the
University.*

> Translator's Note: "Sprache" is the one word used throughout
> — it can be translated as either 'speech' or 'language,' so when
> these words appear, please consider the alternative as well.

My dear venerated guests! The organisers of this
university course have asked me to introduce the reflections
of the day through some remarks and so I will introduce

today's work in a certain aphoristic manner to open our discussion. I am aware that this is no easy task at present. Once in Stuttgart I gave a short course to a smaller circle regarding the items I want to talk about today and it became clear to me that one really needs a lot of time to discuss such controversial things as we would like to talk about today. So I'm only going to suggest a few things about the spirit of our reflection which is required by Anthroposophy in relation to observing human speech.

When speech is the subject and when one sets the goal to treat speech scientifically, then one must be clear that it is not as easy to have speech as an object for scientific treatment as it is for instance about human beings relating to nature or to the physical nature of the human being. In these cases, one has at least a clear outline for the observation of the object. Certainly one can discuss to what a degree observation lies at its foundation, or if it is merely a process being grasped through human research capabilities of an unknown origin. However, this is then a discussion which happens purely within the course of thought. What is presented as an object of observation is a closed object, a given.

This is not the case in spoken language. A large part of speech means that through a person speaking, something is unfolding which was already in the subconscious regions of the human soul life. Something strikes upward from these subconscious regions and what rises, connects to conscious elements which gradually, like harmonics, move with it in an unconscious or subconscious stream. That which is momentarily present in the consciousness, what is present as we speak, that is only partially the actual object essential for our observation. One can, if one remains within the current speech habits of people, acquire a certain possibility of bringing language as an object into consciousness, also when one is speaking. I would like to present in a modest way an example which could perhaps illustrate this.

During Christmas in Dornach I held a lecture cycle at the Goetheanum regarding pedagogical didactic themes. This lecture cycle came about as a request which resulted in a row of English teachers coming to the lectures which they had asked for. When it became known that this course was going to take place, people from other countries in western and middle Europe, namely Switzerland, also gathered to listen to the lectures. Because this course couldn't contain the 900 visitors in the large auditorium of the Goetheanum, but could only be held in a smaller hall, I was notified to give the lectures twice, one after the other. Already before this I believed that to a certain degree it would be necessary to separate the English speakers from those who belonged to other nationalities — not out of political grounds; I stressed this clearly. The lecture cycle was given throughout also for the English speakers; because when people want to hear something about Anthroposophy, wherever it is presented, I always speak German to them. I thought this was something through which its "Germanic" nature could be documented, whereby the German character and German language can be served.

In one of these lectures I had to discuss ethical and moral education. I tried in the course of the lectures to show how the child can be guided in these steps inwardly in its earthly life, which could bring about a certain ethical and moral attitude in the child.

If I would today again speak in front of individuals who listen in the same way as some had listened yesterday, then one could again construe that I spoke out of direct experience, as it happened yesterday, when I spoke about the Trinity. However, Dr Rittelmeyer responded so clearly with a comparison between the book and the mind, which understandably I didn't wish to do.

In this lecture I want to indicate the ethical, moral education towards which the child needs to be orientated so that it is done in the right way: feelings of gratitude, interest in the world, love for the world and his or her own activity and action; and I would like to show how, through love imbuing their activity and actions they are steered to something which can be called human duty. It would be necessary for this trinity to be taken directly out of life's experience and express them in three words — we're talking about language here. I arrived at the first two steps, Gratitude and Love, then the third step: Duty. Despite having to give the lecture twice, once from 10 to 11 o'clock for the English audience, and a second time from 11 to 12 for other nationalities, the latter with their frame of mind being that of central Europeans, I actually had to do these lectures which should simply have been parallel, in quite a different way for the English than for the Germans because I needed to make an effort to live into the mood of my audience. Something similar applied to the other days but on this day, it was particularly necessary.

Why was this so? Yes, while I spoke about duty during the hour from 11 to 12, my entire audience experienced it through words of the German language; I had spoken in the first hour from 10 to 11 what I had to say about their experience of the "Pflicht"-impulse, which they call "duty." Now it is quite a different experience when one expresses the word "Pflicht" to the word "duty" and in the 11 to 12 o'clock lecture I had to allow nuances of experience to flow into what happens when one says "Pflicht." When one says "Pflicht" one touches an impulse through these words which comes out of the emotional life, which flows directly into experience as something — which I want to say verbatim — is related to "pflegen" (to care for). Out of this activity flows the feeling, as to what belongs to this activity. This is the impulse which one designates to the word "Pflicht." Something quite different lives in the soul when this impulse is designated by the word "duty," because just as much as the word "Pflicht" points to the feelings, so the word "duty" points to the intellect, to the mind, to what is directed from within, like how thoughts are being conducted when one goes over into activity. One could say "Pflicht" is fulfilled through inner love and devotion, duty is fulfilled from the basis of a human being, when sensing his human dignity, must say to himself: you must obey a law which penetrates you, you must devote yourself to the law which you have grasped intellectually. This is roughly characterised. However, with this I want to bring into expression how inner complexes of experience are quite different between one word and another, and yet despite this the dictionary says the German word "Pflicht" translates to the English word of "duty". This is however transmitted by the spirit of the folk, in the folk soul and in the speech, you have nuances of the entire folk soul. You are going to see that in the soul of central Europeans, in relation to this, it looks quite different compared with souls of other nationalities; that the soul life is experienced quite differently in speech by central Europeans compared with the English nation.

A person who has no sense for the unconscious depths of soul where speech comes from, which lies deeper than what is experienced consciously, will actually be unable to obtain a sober objectivity for scientific observation of speech. One should be clear about one thing. With nature observation the objects present themselves, or one can clean them up through outer handling in order to have the object outside oneself and thus able to research it. To consider speech it is necessary to first examine the process of consciousness in order to come to what the object essentially is which one wants to examine. So one can, where speech is the subject, not merely consider what lives in human consciousness, but in considering speech one needs to have the entire living person before you who expresses himself in speaking and speech.

This preparation for the scientific speech observation is very rarely done. If such preparation would be undertaken then one would, if one takes linguistic history or comparative linguistics, move towards having a deep need to first contemplate the inner unconscious content of that language, the unconscious substance which in speaking only partly comes to expression.

Now we arrive at something else, namely, during the various stages of human development this degree of consciousness associated with language was quite varied. It was quite different for example during the times in which Sanskrit had its origins; different again during the time the Greek language developed, another time than we had here in Germany — but here nuances became gradually less recognisable — and in another time, it happened for instance in England. There are already great variations in the inner experience of the conduct in the English language when used by an Englishman or American, if I observe only the larger differences. Whoever takes up the study of dialects will enter into how the different dialects in the language is experienced by the people who use it and take note of all the complicated soul impulses streaming through it which comes into expression as speech in the vocal organism. It is for instance not pointless that when the Greek speakers say "speech" (Sprache) or when they say "reason" (Vernunft), they consider both these words as essentially the same and can condense them into one word, because the experience within the words and the experience within thoughts, within mental images, flow together, undifferentiated, in the Greek application of speech, while in our current epoch differentiations show themselves in this regard. The Greek always felt words themselves rolled around in his mind when he spoke; for him thoughts were the "soul" and words streaming in formed the "body", the outer garments one could call it, the word-soul streaming in thought. Today we feel, when we clearly bring this process into consciousness, as if on the one side we would say a word — the word streams towards what we express — and on the other side the thoughts swim in the stream of words; it is however soon clearly differentiated from the stream of words.

If we return for instance to Sanskrit then it is necessary to undergo essential psychological processes first, to experience psychic processes, in order to reach the possibility to live inwardly with what at the time of Sanskrit's origin was living in the words. We may not at any stage confront Sanskrit with the same feelings when regarding its expression, when regarding its language, as we would do with a language today.

Let's take for example a familiar word: "manas". If you now open the dictionary you would find a multitude of words for "manas": spirit, mind, mindset, sometimes also anger, zeal and so on. Basically, with such a translation one arrives at an experience of a word which once upon a time existed when it was quite clearly and inwardly experienced, not nearly. Within the epoch when Sanskrit lived at the height of its vitality, with a different soul constitution as it has today, it was essentially something different. We must clearly understand that human evolution already existed as a deep transformation of the human soul constitution. I have repetitively characterized this transformation as having taken place somewhere in the 15th Century. There are however ever and again such boundaries of the epochs when going through human evolution, and only when one can follow history as the inner soul life of the people can one discover what really existed and how the life of speech played its part.

It was during such a time when the word "manas" could still be grasped inwardly in a vital way, when something existed which I would like to call the experience of the meaning of sound. In an unbelievable intense way one experienced what lived inwardly in the sounds, which we designate today as m, as a, as n and as s. The life of soul rose to a higher level — still dreamily, yet in a conscious dream — with its inward living within the organism when the vocals and consonants were pronounced. Whoever uses *such* scientific tools for researching how speech lives within people, will find that everything resembling consonants depends upon people placing themselves into external processes, into things, and that the inner life of things with their own inner, but restrained gestures, want to copy it. Consonants are restrained gestures, gestures not becoming visible but which through their content certainly capture that which can outwardly be experienced in the role of thunder, lightning flashes, in the rolling wind and so on. An inner inclusion of oneself in outer things is available when consonants are experienced.

We actually want to, if I might express myself like this, imitate through gestures all that lives and weaves outside of us; but we restrain our gestures and they transform themselves within us and this transformation appears as consonants.

By contrast, by opposing external nature, mankind has living within itself a number of sympathies and antipathies. These sympathies and antipathies within their most inner existence form gestures out of the collective vowel system, so that the human being, through experiencing speech, lives in such a way that he, within the nature of the consonants, imitate the outer world — but in a transformed way — so that in contrast, through the vowels, he forms his own inner relationship to the outer world.

This is something which can certainly be understood and examined through today's soul life if one enters into the concrete facts of the speech experience. It deals with what is illustrated as imagination, not as some or other fantasy, but that for example the inner process of the speech experience can really be looked at.

Now in ancient times, in which Sanskrit had its original source, there was still something like a dreamlike imagination living within the human soul. Not a clearly delineated mental picture like we have today was part of man, but a life in pictures, in imaginations — certainly not the kind of imaginations we talk about in Anthroposophy today, which are fully conscious with our sharply outlined concepts, but dreamlike instinctive imaginations. Still, these dreamlike imaginations worked as a power. If we go back up to the time we are talking about, one can say these imaginations lived as a vital power in people: they sensed it, like they sensed hunger and thirst, only in a gentler manner. One painted in an internal manner, which is not painting as in today's sense, but in such a way as to experience the inward application of vocalisation, like we apply colour to a surface. Then one lives into the consonants through the vocalization, just as when, by placing one colour beside another, one brings about boundaries and contours. It is an inner re-experience of imaginations, which presents an objective re-living of outer nature. It is the re-living of dreamlike imaginations. One surrenders oneself to these imaginations and inverts the inner processed imaginations through the speech organs into words.

Only in this way does one imagine the inner process of the life of speech in the way it was once experienced in human evolution. If one becomes serious about such an observation, for example through the experience of tones, which we call 'm' today, we notice that with the experience of this sound, we stand at once on the boundary between what is consonant and what is vowel. Just like we paint a picture and then the colours, which have their inner boundaries and outer limitations and do not continue over the surface, just so something is expressed in the word "manas". With 'a' something resembling human inwardness is sensed. If one wishes to describe the word "manas" I have to say: In olden times people lived in their dream-like imaginations in the language, just as we experience speech consciously now. We no longer live in relation to speech in dream pictures, but our consciousness lies over speech. Old dreamlike imaginations flowed continuously in the language. So, when they said the word "manas" they felt as if in some kind of shell, they felt their physical human body in as far as it is liquid aqueous, like a kind of shell, and the rest of the body as if carried in a kind of air body. All of this was experienced in a dreamlike manner in olden times when the word "manas" was spoken out. People didn't feel like we do today in our soul life, because people felt themselves to be the bearers of the soul life — and the soul itself one experienced as having been born out of the supersensible and super-human forces of the shell.

You must first make this experience lively if you want to understand the content of older words. We must realise that when we experience our "I" today it is quite different from what it was when the word "ego" was for instance come across in humanity in earlier times, when the word "aham" was experienced in the Sanskrit language. We sense our "I" today as something which is completely drawn to a single point, a central point to which our inner being and all our soul forces relate.

This experience does not underlie the older revelations of the I-concept. In these olden times a person felt his own I as something which had to be carried; one didn't feel as if you were within it. One then experienced the I to some extent as a surging of soul life swimming independently. What one felt was not indicated by the linguistic context — what lay in the Sanskrit word "aham" shows it is something around the I, which carries the I . While we feel the I inwardly as will impulses — we really experience it this way today — which permeates our inner being, we say that as its central point it is a spring of warmth, which streams with warmth — to make a comparison — streaming out on all sides, this is how the Greek or even the Latin experienced the I like a sphere of water, with air permeating this sphere completely. It is something quite different to feel yourself living in a sphere of water within extended air, or to experience the inward streaming towards a central point of warmth and to stream out warmth to the periphery of the sphere and then — if I might use this comparison more precisely — to be grasped as a sphere of light.

These are all symbols. Yet the words of a language are in this sense also symbols, and if you deny the ability of words to indicate symbols, you would be totally unable to be impressed by such a consideration. It is necessary in the research of linguistics that one first lives into what actually has to become the object of linguistics. Now, one finds that in ancient times, the language had a considerably different character than what exists in civilisation's current language; further, one finds that the physical, the bodily, played a far greater part in the establishment of phonetics, in the establishment of word configuration. The human being gave much more of his inner life in speech. That is why you have 'm' at the start of "manas" because this enclosed the human being, formed a contour around him or her.

When you have Sanskrit terms in front of yourself, you soon notice you can experience the nature of the consonants and vowels within it. You notice how in this activity an inner experience in the external events and external things are present and how this results in the consonants being imitated, so vocal sympathies and antipathies are discovered where the word process and the speech process merge. In ancient times a much more bodily nuance came about. One had a far greater experience in the ancient life of speech. This one can still experience. If today you hear someone speaking in Sanskrit or the language of an oriental civilisation, how it sounds out of their bodily nature, and how speech absorbs the musical characteristics, it is because such an experience rises out of the musical element. Only in a later phase of human evolution the musical elements in speech split away from the logical, thus also away from the soul life, into mere conceptions.

This is still noticeable today. When for instance you compare the inner experience in the German and in the English language, you notice that in the English language the process of abstract-imagery-life have made greater progress. If we want to live in the German language today we must live into those forms of the speech which came about in New High German. [*'Hochdeutch'* or High German *is the pure German language without the influence of dialects, which is also understood by most Germans*. New High German *differs from* Old High German *as the latter refers to more historic times*. Schriftliche Deutch *is the German most widely used in school instruction, standardization, etc. — translator*.] The dialects still lets our soul become immersed in a far more intensive and vital experience. The actual spiritual experience of the language is primarily only possible in High German. Thus, a figure such as Hegel who was born out of this spirit, for whom the mental images are particular to him and yet it is also quite connected to a particular element within the language, out of these causes it has come about that Hegel is in reality not translatable into a western language, because here one experiences the literal fluency (Sprachliche) even more directly.

When you go towards the west you notice throughout within the observation how the soul unfolds when it is given over to the use of language: the soul experiences it intensively, however the literal fluency (Sprachliche) is thrown out of the direct soul experience throughout; it flows away in the stream of speech and continuously, to some degree, out of the flowing water something is created like ice floes, like when something more solid is rolling over the waves — as for instance in English. When, by contrast, we speak High German, we can observe how a person in the stream of speech is in any case within the fluidity of it but in which there are not yet any ice blocks which have already fallen out of the literal fluency, which are connected with the soul-spiritual of the human being.

Now when we come towards the east, one finds this process in a stage which is even further back. Now you don't see ice floes which are thrown out of the stream of speech, and which are not firmly connected with it; here also, as not in High German, the entire adequacy of thoughts are experienced with the word but the word is experienced in such a way that a person retains it in his organism, while thoughts in their turn flow into the words, which one runs after but which actually goes before you.

These are the things which one has to live through when one wants to really understand literal fluency. One can't experience this if one doesn't at least to a certain degree take on the contemplation which <u>Goethe</u> developed for the observation of the living plant world and which, when in one's inner life, these are followed with inner consequential exercises, leading towards mental pictures about what is meant in Anthroposophy. Anyway, if you want to look at the language, you must observe it in such a way that you live within the inner metamorphosis of the organising of the language, experience in its inner concreteness, because only then will you have in front of you, what the speech process is. As long as you are unable to rise up to such inner observations of speech, you are only looking at speech in an outer way, and you will be unable to penetrate the actual living object of language. As a result, all kinds of theories of speech have appeared. Ideas about language have in many cases become thought-related regarding the origins of language; a number of theories have resulted from this. _Wilhelm Wundt_ enumerated them in his theory of language and picked them apart critically.

This is the way things are today in many areas and how it was observed yesterday. When the bearers of some scientific angle today raises into full contemplation regarding what he has observed within the science and he represents it thus, then talk starts to develop about "decline". This is actually not really what Anthroposophy wants to tell you. Basically, for example, yesterday very little was said about decline; but very much not so in the case of those who stand within theology, for they are experiencing a decline.

Similarly, there is also talk regarding the philosophy of language, of declining theories, for instance with the "theory of creative synthesis/invention" (Erfindungstheorie). _Wundt_ lists his different theories. Following on the theory of invention the language developed in such a way that humanity, to some extent, fixed the designations of things; however, this is no longer appropriate for current humanity because today the question they ask is how could the dumb have fixed forms of language while still so primitive?

As his second, Wundt presents his "theory of wonder" (Wundertheorie) which assumes that at a certain stage of evolution human speech/language arrived as a gift from the Creator. Dr Geyer already dealt with this yesterday; currently it is no longer valid for a decent scientist to believe in wonder; it is prohibited, and so the theory of wonder is no longer acceptable. Further down his list is the "theory of imitation" (Nachahmungstheorie) which already contains elements which have a partial authorisation because it is based on elements of consonants in speech being far more on an inner process than what is usually imagined. Then the "natural sound theory" (Naturlauttheorie) followed which claimed that out of inner experience the human being aspired towards phonetically relating what he perceived out in nature, into the form of speech, according to his sympathies or antipathies. These theories could be defined differently. Today it is quite possible to show that on the basis of those who criticise these theories, it becomes apparent that these theories can't determine the actual object of language.

Dear friends, the thing is actually like this: Anthroposophy — even when people say they don't need to wait for her — can still show in a certain relationship, what can be useful in this case, through which — even in such areas as linguistics — firstly the sober, pure object is to be found, on which the observation can be based.

Obviously anything possible can be discussed, also regarding language, even when one actually doesn't approach it as a really pure object. Anthroposophy bears within it a profound scientific character which assumes that first of all one must be clear what kind of reality there is to be found in specific areas, in order for the relationships we have regarding truth and wisdom to penetrate these areas, so that these areas of reality can actually become inward experiences. As we saw happening here yesterday, then in relation to such earnest work which is not more easily phrased in other sciences, it is said that these Anthroposophists stick their noses into everything possible, then it must be answered: Certainly it is apparent that Anthroposophy in the course of its evolution must stick its nose into everything. When this remark doesn't remain in superficiality, this 'Anthroposophy sticks her nose into everything possible' — but if one wants to make progress to really behold and earnestly study the results, when it comes down to Anthroposophy sticking its nose into everything, only then, when this second stage in the relationships to Anthroposophy is accomplished, will it show how fruitful Anthroposophy is and in how far its legitimacy goes against the condemnation that it merely originates from superficial observation!

ON-LINE
RESOURCES

- Rudolf Steiner Archive & e.Lib. *The Impulse of Renewal for Culture and Science*, Bn/GA# 81, translated by Hanna von Maltitz.
- Rudolf Steiner Archive & e.Lib. *Erneuerungs-Impulse für Kultur und Wissenschaft*, Bn/GA# 81, original German.
- Rudolf Steiner Archive & e.Lib. *The Impulse of Renewal for Culture and Science*, Bn/GA# 81, Side-by-Side Compare.
- Fine Art Presentations – e.Gallery. *Hanna von Maltitz* (see more of Hanna's paintings)
- Rudolf Steiner Archive & e.Lib.

Other on-line links

The Philosophy of Freedom/Spiritual Activity:

www.rsarchive.org/Books/GA004/index.php

This book, written in 1894, is a fundamental treatment of Steiner's philosophical outlook. Together with *Truth and Science* and *The Riddles of Philosophy* it might be considered as one third of a philosophical trilogy. The emphasis of the book is not on "freedom" as ordinarily understood but is on "freiheit" or what might be termed freeness, or freehood. This freeness is a self-starting, self-directing activity of a spiritual sort. Here, the term "spiritual" is used in a sense not incompatible with the use of the word in the tradition of the German idealistic school of philosophy. The *Philosophy of Freedom* was originally published in German as, *Die Philosophie der Freiheit. Grundzüge einer modernen Weltanschauung. Seelische Beobachtungsrelultate nach naturwissenschaftlicher Methode.*